Foreword

The project leading to this report is part of a CIRIA programme 'Concrete Techniques - Site Operations', and was carried out under contract by Taywood Engineering Limited.

Research Team

Dr P B Bamforth	–	Taywood Engineering Limited
Dr W F Price	–	Taywood Engineering Limited

Steering Group

The project was carried out and the report prepared under the guidance of the following Steering Group:

Mr R A McClelland (Chairman)	–	Alfred McAlpine Construction Ltd
Mr R Cather	–	Arup Research and Development
Mr B Cowling	–	Appleby Group
Mr S Crompton	–	Ready Mixed Concrete (UK) Ltd
Dr T A Harrison	–	British Ready Mixed Concrete Association
Dr B Marsh	–	Building Research Establishment
Mr G G T Masterton	–	Babtie Shaw & Morton
Mr A McGibney	–	Civil and Marine
Dr J B Newman	–	Imperial College of Science, Technology and Medicine
Mr A J Nicklinson	–	Trafalgar House Construction (Major Projects) Ltd
Mr P Owens	–	Quality Ash Association
Mr R Roberts	–	Concrete Advisory Service
Mr P Titman	–	Edmund Nuttall
Mr C Turton	–	Design Group Partnership
Dr B W Staynes	–	CIRIA Research Manager

Funding

The project was funded by the Department of the Environment and the Institution of Civil Engineers R&D Enabling Fund.

Report 135

1995

Concreting deep lifts and large volume pours

P B Bamforth
W F Price

CIRIA

CONSTRUCTION INDUSTRY RESEARCH AND INFORMATION ASSOCIATION
6 Storey's Gate, Westminster, London SW1P 3AU
Tel 0171-222 8891   Fax 0171-222 1708

Summary

Recommendations are given for the construction of deep lifts and large volume pours. The intention is to make the designer, the contractor and the concrete producer, aware of the particular features of these construction methods, and to provide guidance on how to avoid potential problems.

The topics covered include specification and design; concrete mix selection in relation to critical properties; placing methods; construction sequence; procedures for minimising defects; and quality assurance, monitoring and compliance testing. Checklists for planning deep lifts and large volume pours are included.

P B Bamforth, W F Price
Concreting deep lifts and large volume pours
Construction Industry Research and Information Association
Report 135, 1995

CIRIA ISBN	0 86017 420 4
Thomas Telford ISBN	07277 2025 2
ISSN	0305 408X

© CIRIA 1995

Keywords	Reader Interest	Classification	
Concrete, deep lifts, large volume pours, thermal cracking, concrete mix design, formwork, compaction, low heat concrete	Design, specification, construction and supervising engineers involved in heavy civil engineering works.	AVAILABILITY	Unrestricted
		CONTENT	Review of available guidance
		STATUS	Committee guided
		USER	Civil engineers

Health and Safety

Construction activities, particularly on construction sites, have significant Health and Safety implications. These can be the result of the activities themselves, or can arise from the nature of the materials and chemicals used in construction. This report does not endeavour to give comprehensive coverage of the Health and Safety issues relevant to the subject it covers, although specific points are mentioned where appropriate. Other published guidance on specific Health and Safety issues in construction should be consulted as necessary.

Contents

FIGURES

TABLES

Notation

A_n Effective area of newly cast concrete
A_o Effective area of mature concrete
e_r Restrained strain in concrete
e_{sh} Drying shrinkage strain
e_{th} Total thermal strain
E_n Modulus of elasticity of newly cast concrete
E_o Modulus of elasticity of mature concrete
f_b Average steel-concrete bond strength
f_{ct} Concrete strength in tension
K Restraint multiplier for sustained load and creep
M_a Mass of aggregate
M_c Mass of cement
M_i Mass of ice
M_w Mass of water
R Restraint factor
ΔT Temperature change in concrete
T_i Difference between the centreline peak temperature and ambient temperature
T_a Temperature of aggregate
T_c Temperature of cement
T_w Temperature of water
α_c Coefficient of thermal expansion of concrete
γ_m Materials safety factors
ρ Percentage of steel
ϕ Bar diameter
ε_{tsc} Tensile Strain Capacity

Abbreviations

ggbs   Ground Granulated Blastfurnace Slag
pfa   Pulverised Fuel Ash
NDT   Non Destructive Testing
c.s.a.   Cross Sectional Area

Glossary

Air entrainment	The intentional incorporation into a concrete mix, by means of an admixture, of minute discrete bubbles of air typically less than 0.3mm (300 microns).
Autogenous healing	The natural sealing of a crack in concrete by siltation together with further hydration of cement or by the deposition of calcium hydroxide with consequent conversion to calcium carbonate.
Bleeding	The phenomenon of water migrating to the surface of fresh concrete due to settlement of particulate materials.
Blended cement	See Composite cement.
Cohesiveness	The degree to which fresh concrete resists segregation.
Cold joint	Plane of weakness caused by interruptions to placing which results from new concrete being cast against unprepared stiffened concrete.
Cold weather concreting	Placing and finishing concrete when air temperatures are below 5°C or when freezing is likely to occur soon after concreting.
Compactibility	The ease with which the entrapped air in concrete can be expelled.
Composite cement	A cement or combination containing two or more main constituents, one of which is Portland cement, mixed at a cement works or at a ready-mix plant or on site.
Consistence	The resistance to flow of fresh concrete.
Core	The innermost part of a (thick) concrete section which is generally unaffected by external ambient temperatures.
Creep	The continuing deformation of concrete under sustained stress.
Curing	The action of preventing or minimising the loss of moisture from fresh concrete, and maintaining a satisfactory temperature so that desired properties may develop.
Drying shrinkage	The long-term reduction in size of a concrete element caused by the loss of water upon drying.
Early age thermal contraction	The reduction in length of a concrete element caused by its cooling from the elevated temperatures developed during hydration of the cement.
Finishing	Mechanical operations carried out on the surface of placed concrete to achieve the required texture, appearance and serviceability.

Fresh concrete	Concrete which has insufficient reaction between the cement and water to exhibit significant mechanical properties.
Hardened concrete	Concrete which exhibits mechanical properties as a result of the reaction of cement and water.
Heat of hydration	Heat generated by the chemical reaction of cement and water.
Hot weather concreting	Placing and finishing concrete when air temperatures are above 25°C and the combination of temperature and humidity is such that the quality of fresh or hardened concrete may be impaired unless preventive measures are taken.
Joint-construction	A monolithic joint with no provision for movement.
Joint-movement	A joint formed or induced to allow movement to occur with one or more degrees of freedom.
Laitance	A layer of cement and fine sand with excess water content on top of concrete.
Matrix	The part of concrete which is composed of cement paste and sand.
Mobility	The ease with which fresh concrete can be made to flow.
Placing	All operations necessary for introducing the fresh concrete into formwork or other enclosures.
Plastic concrete	See Fresh concrete.
Plastic settlement cracks	Cracks which form due to differential settlement of concrete while still in a fresh (plastic) state.
Plastic shrinkage cracks	Cracks which form because of excessive loss of moisture from concrete while still in a fresh (plastic) state.
Restraint	Anything (internal or external) which prevents the free movement of concrete.
Retempering	Addition of water and remixing of concrete to increase its workability.
Rheological	Pertaining to the flow of concrete.
Segregation	Loss of uniformity of fresh concrete mix due to separation of one or more of the constituents of the mix.
Settlement	Sedimentation or the act of the heavy solid particles settling in fresh concrete.
Stability	See Cohesiveness.

Stiffening time	Time beyond which reworking of concrete would be detrimental.
Tensile strain capacity	The tensile strain at which a concrete will crack.
Transporting	The conveying of fresh concrete from the point of discharge from the mixer to the point of discharge into the formwork or other enclosure.
Warping	The distortion of a slab caused by temperature or moisture gradients across the depth of section.
Workability	The rheological properties of fresh concrete influencing the ease with which it can be placed, compacted and finished.

1 Introduction

1.1 BACKGROUND

This report supersedes two earlier CIRIA reports, numbers 15 and 49. Report 15, *Placing concrete in deep lifts*[1], was published in 1969, and Report 49, *Large concrete pours – a survey of current practice*[2], was published in 1974.

Since that time there have been significant changes in concrete construction, in relation to both the available plant and the materials used. Combined with the fact that there has been no evidence of significant defects arising from placing concrete in deep lifts or large volumes when appropriate materials and procedures have been adopted, this has resulted in pours getting taller and more massive in a continuing effort to build more cost-effectively. It is not uncommon for columns and walls to be cast in single lifts of 10 m or more, and in foundations and other massive elements, single pours have routinely exceeded 1000 m^3.

This report provides guidance to the designer, the contractor and the concrete supplier, based on the experience of many deep lifts and large volume pours which have been successfully executed in the UK and overseas. Combined with the advancing knowledge of the performance of concrete in construction, the wider use of composite cements and chemical admixtures, and the way in which the concrete constituents influence the behaviour of fresh and hardened concrete, this has led to increased confidence in applying these techniques in practice.

Present experience and knowledge is presented in the form of technical recommendations, backed by checklists to ensure that these and other appropriate measures are taken into account when designing and placing concrete in deep lifts or large volumes.

The report is in sections to provide easy access to information. Section 2 relates specifically to the role of the designer, in particular in relation to specification and the control of cracking. Section 3 covers planning and represents a comprehensive pre-construction check list for the contractor. This should be read in conjunction with Section 4 which covers concrete mix design, and how to achieve the performance characteristics of the fresh and hardened concrete appropriate to the method of construction, and either Section 5 or Section 6, which provide guidance on construction methods for deep lifts and large volume pours respectively.

Section 7 defines the advantages of deep lifts and large volume pours, reviews the main technical requirements for successful construction, and gives examples.

1.2 DEFINITIONS

Deep lifts - Within the context of this guide, a deep lift is considered to be one which is tall in relation to its minimum dimension, i.e. a wall or a column. Thus, the principal consideration for the construction of a deep lift is access and how this affects delivery and compaction of the concrete. Concrete must be placed and properly compacted into a relatively confined space which cannot always be observed clearly from the top of the form. Hence a deep lift can be defined as one which demands special attention to logistical and technical considerations, such as:

- concrete distribution to the point of placement
- segregation
- compaction
- plastic settlement
- surface finish

Other technical considerations, such as concrete supply, casting sequence and heat of hydration, may also be significant, but are not specific to deep lifts. The listed considerations are also relevant in general concrete construction, but are of increased importance when the decision is taken to cast deep slender elements.

Large volume pours - Concrete Society Digest No. 2[3] defines a mass pour as one of sufficient size to demand special attention to logistical and technical considerations, such as:

- concrete supply
- casting sequence
- cold joints
- plastic settlement
- heat of hydration
- early age thermal cracking

The American Concrete Institute Manual of Concrete Practice 207[4] defines mass concrete as 'Any volume of concrete large enough to require measures to be taken to cope with the generation of heat and attendant volume change to minimise cracking'. In the latter definition, the practical aspects of large volume placements are not considered to warrant special attention, probably due to the fact that, in the US, large volume pours have been commonplace for many years, and practice is well established. In CIRIA Report 49, of twenty one large pours investigated in the UK in 1974, only two exceeded 1000 m^3, while in the US pours in excess of 6000 m^3 had been completed as early as 1968[5].

It is clearly inappropriate to give a specific quantitative definition of a 'large volume' pour, as this will be relative to the experience of the construction team and the scale of operations on a particular site. It is sufficient to stress once again the important features defined above which relate to continuity of casting and avoidance of defects.

1.3 BENEFITS

The principal benefits of deep lifts and large volume pours are the savings in cost and time resulting from the reduction in the number of joints. These are expensive to form, requiring stop ends when vertical and careful preparation of the concrete surface, regardless of orientation, when new concrete is cast against old.

In a discussion of the relative disadvantages of construction joints, CIRIA Report 49 concluded that:

'The most significant disadvantages of construction joints in large sections are likely to be those of:

1. Delays in construction progress with consequential costs
2. Labour and material costs involved in making the joints
3. Practical difficulties in forming fully-effective joints.'

The current trend towards the imposition of very high stresses on massive concrete sections (e.g. heavily reinforced foundations to core-shaft tower blocks) has tended to create a situation where the formation of sound construction joints or temporary stop-ends is sometimes virtually impossible because of very congested reinforcement. Similar conditions are often found in large sections such as bridge decks, where the position of void formers, stressing cables, reinforcement, ducts, etc. render the installation of construction joints extremely difficult.

The disadvantages of cracks which might occur in situations where construction joints are not used, appear to be comparatively minor. The importance of such cracks should be considered in relation to the structural requirements of the section, together with considerations of durability. For the protection of reinforcement, a maximum crack width of 0.30 mm has been considered acceptable.

In addition, the elimination of joints removes potential cracks and zones of weakness, which are often highlighted in water-retaining structures.

1.4 REVIEW OF TECHNICAL CONSIDERATIONS

While there can be significant benefits of scale resulting from the programmed use of deep lifts and large volume pours, their successful execution demands an awareness of the specific technical requirements of the concrete in relation to the geometry of the element and the proposed methods for transportation, placing and compaction.

For **deep lifts**, the geometry demands that the following are given special attention.

- Uninterrupted delivery of the concrete into the formwork.
- Avoiding placing concrete by free fall from the top of the shutter, particularly at the start of a pour. This can lead to honeycombing, blowholes and sand runs in the most severe cases.
- Control of the pattern of placing avoiding peaks and troughs. This may require the use of formwork access doors.
- Achievement of adequate compaction. In very deep lifts, it is difficult to control the location of poker vibrators from the top. Consideration must, therefore, be given to appropriate means of locating and controlling vibration and the provision of lighting if visibility is poor.
- Settlement, and consequential cracking. This may be most severe in very slender elements, or splayed columns, where bridging may result in horizontal 'tears' as settlement is restrained.

- General surface finish and defects. The type of formwork, the form release agent used, the concrete mix type and its homogeneity, and the method of delivery and compaction will all influence surface finish. When a high quality finish is required, therefore, careful attention must be paid to all these factors, and their control during the construction process. While not specific to deep lifts, these factors may be more critical when access is difficult.

For **large volume pours**, the geometry has an equally significant influence on the materials and method which must be employed. However, in this case, the problems relate more specifically to the magnitude of the volume of concrete to be placed, and the achievement of a monolithic and homogenous element.

Particular attention must, therefore, be given to:

- Planning to achieve continuity of concrete supply and compatibility of rates of delivery, distribution, compaction and finishing.
- Avoidance of cold joints. Here the rate and sequence of placing must be considered in relation to the stiffening time of the concrete.
- Plastic settlement. Large volume pours are often deep slabs, e.g. raft foundations, and plastic settlement may result in cracking, characteristically above the top mat of steel.
- Early age thermal cracking. Excessive thermal gradients, or restraint to bulk thermal contraction can cause cracking within days, or weeks, of casting. Cracking can be controlled by the use of reinforcement, and the risk of cracking can be minimised by appropriate materials selection and mix design, control of the mix temperature, insulation or cooling of the cast element and planning the sequence and timescale of construction.

CIRIA Report 49 concluded that '...in general, no significant adverse effects have been caused by the placing of concrete in large pours for reinforced concrete structures...' and experience in the intervening years, with ever increasing pour sizes – the maximum reported[6] is 17 000 m^3 cast in Frankfurt, Germany, over a period of 78 hours – has provided no evidence to change this conclusion.

Similarly, deep lifts have been increasing in size, demonstrating their successful use when properly designed and executed.

Where difficulties do arise, this is usually due to a lack of awareness of those aspects of the concrete and the construction method which demand special attention, namely:

For deep lifts:

- concrete placement and compaction
- plastic settlement
- surface finish, including specified finishes

For large volume pours:

- continuity of concrete supply
- cold joints
- plastic settlement
- early age thermal cracking

These technical aspects are addressed in detail in the relevant sections of the report.

2 The designer's role

This section provides guidance to the designer, in particular in relation to the impact of the specification on construction and to the control of early age thermal cracking.

2.1 APPROACHES TO THE SPECIFICATION

In general there are two basic approaches to specification – performance and prescriptive – although in practice, they are likely to be a combination of the two. In the former case the designer specifies performance limits and allows the contractor the flexibility to meet these by his own endeavours. When using this approach the designer must ensure that all aspects of the specification are compatible. For example, a minimum cement content for durability may be inconsistent with low heat of hydration unless either a specific cement type is used, or special construction measures are employed, such as cooling of the fresh concrete or use of embedded cooling pipes. Such cases should be highlighted within the contract documents and guidance should be provided as appropriate. In addition, the designer should define the assumptions which have been made in relation to the concrete mix type and the construction programme. This is particularly important in relation to limits on early age thermal cracking, where assumptions are made on:

- Temperatures and temperature differentials
- Coefficient of thermal expansion of the concrete
- Strain capacity of the concrete
- Restraint to thermal movement at critical locations (which may also involve assumptions about pour sizes and casting sequences)

This performance approach will require close cooperation between designer and contractor.

The prescriptive approach involves the designer in more detailed specification of the concrete mix type, maximum pour sizes and the construction sequence. In this case, the designer has responsibility for the completed structure, provided that accepted good practice is exercised during the construction process.

When using either of these approaches it may be necessary to undertake either pre-construction trials or monitoring to establish compliance. Programme advantages may be achieved, however, if the designer is willing to waive the necessity for trials or in-situ measurements by preparing a functional specification, thus giving the contractor options to maximise the benefits of particular materials and methods.

As in all construction projects, the designer can have significant impact on the construction process. It is, therefore, essential that he either makes himself aware of that process or that he works closely with the contractor. This is particularly important with regard to design for control of cracking. In the case of deep lifts and large volume pours, the designer can influence the construction process by placing limits on:

1. The temperatures and temperature differentials resulting from heat of hydration
2. Pour sizes and configuration and the location of joints

3. The sequence of construction and the timescale (e.g. delay between adjacent elements)
4. On concrete mix constituents and mix proportions.

In addition there may be structural requirements which demand quantities and distribution of steel which lead to construction difficulties.

In each case, the designer must be aware of the impact of the design on the construction process – for example, in heavily congested areas. Where such situations are unavoidable due to some specific serviceability requirement, the designer should highlight these in the contract documents. Limits on pour sizes or the use of special concrete mixes should also be clearly defined to ensure that the contractor can take appropriate measures.

With regard to the concrete mix, it is important to achieve a balance between specified strength, durability, heat of hydration and the requirements for placing and compaction. In such cases the strength and minimum cement content should not be over specified. For large volume pours in particular, there may be the opportunity to use low strength mixes.

2.2 RESPONSE TO CONTRACTOR'S PROPOSALS

Where the contractor proposes deep lifts or large volume pours, either at the tender stage or after the start of construction, the designer's role will be to review the implications for the structure in two respects:

1. Does the contractor have the experience and expertise to complete the works successfully?
2. Will the functional requirements of the structure still be met?

The designer may judge the expertise of the contractor from the quality of the proposal and his experience of similar construction methods. The assessment of the performance of the structure will require comparison between the design assumptions, e.g. regarding pour sizes, temperature rise, restraint, etc. and revised estimates taking account of the proposed changes. Once again, close liaison between designer and contractor is required. Approval is likely to be given more rapidly if the contractor presents a comprehensive technical proposal, based not only on the practicalities of construction, but also on those matters which are likely to be of greatest concern to the designer, and in particular the risk of cracking.

2.3 EARLY AGE THERMAL CRACKING

In addition to the normal limit states for serviceability and ultimate load, the designer may also need to consider early age thermal cracking.

Reinforced concrete is designed to crack. Where cracking is totally unacceptable prestressing should be employed. Hence the designer's role is to control the location of cracks and to limit crack widths to levels which are consistent with:

- maintenance of structural integrity
- durability

- serviceability
- visual appearance.

CIRIA Report 91, *Early age thermal crack control in concrete*[7], provides guidance on the significance of cracking under the above headings, as shown in Table 1.

Furthermore, the designer must consider whether early age thermal cracks will be additive to cracking which occurs due to service loads and, in the longer term, drying shrinkage. This will depend on the type of element, and its in-service condition. For example, surface cracks which occur in a large volume pour within the first few days after casting, will close up as the bulk of the pour cools down over a longer period, and subsequent drying shrinkage will be insignificant. However, in a deep lift, cracks which occur due to base restraint as the wall cools down will remain and may increase in size as drying shrinkage continues over a long period.

Table 1 *Limiting crack widths*

Limit state	Limiting crack width (mm)	Comments
Structural Integrity	Up to 0.5	According to loading condition
Durability	0.1-0.4	According to environment
Serviceability (in water retaining structures)	<0.2	For self-healing
Appearance	Up to 0.4	Depends on viewing distance

As a general rule, cracks which result from internal restraint (i.e, temperature differentials within a pour) will tend to close with time as the stresses are redistributed and these early stresses need not be included in the structural design. However, cracks which are caused by external restraint will not close, and may increase in width in the longer term. The associated stresses will be additional to load induced stress.

2.4 LIMITATION OF CRACKING

In accordance with CIRIA Report 91, the maximum crack spacing, S_{max}, and the crack width, w, can be estimated using the following equations:

$$S_{\max} = \frac{f_{ct}}{f_b} \frac{\Phi}{2\rho} \tag{1}$$

$$w = S_{\max}\left[R(e_{th} + e_{sh}) - \frac{\varepsilon_{tsc}}{2} \right] \tag{2}$$

where f_{ct} = tensile strength of the concrete
f_b = bond strength of concrete to reinforcement
Φ = bar diameter
ρ = percentage of steel
R = restraint factor

e_{th} = thermal strain = $\alpha_c T_1$

α_c = thermal expansion coefficient

T_1 = difference between the centreline peak temperature and the ambient mean temperature

e_{sh} = drying shrinkage strain

ε_{tsc} = tensile strain capacity

The above equation gives the maximum 'average' crack width, i.e. it assumes that if all the cracks are of the same width, there is a very low risk of the crack width exceeding the calculated value. However, recognising the in-situ variability of concrete, and the fact that the equation assumes values for both the tensile strength of the concrete and its bond strength to the steel, there is a probability that some individual cracks will be greater than the calculated value, while others will be smaller. Conformance must, therefore, be based on an average value taken over the full length of a particular pour.

The contractor will have little influence over many of the above factors but, in a performance specification, will select the concrete mix within the requirements for strength, durability and early age thermal behaviour.

To control the extent of cracking, it is common to specify allowable limits on the centreline peak temperature, T_p, and on temperature differentials, ΔT_{max}, during the post construction period. Typical limits may be specified as follows:

- The maximum temperature at any point within the pour shall not exceed ..... [commonly 70°C]
- The maximum temperature differential within a single pour shall not exceed ... [commonly 20°C]
- The maximum value of mean temperatures between adjacent elements cast at the same time shall not exceed .... [commonly 20°C]
- The maximum value of mean temperatures between adjacent elements cast at different times shall not exceed .... [commonly 15°C].

This is a very simplistic approach, as the object is to limit restrained (or locked-in) thermal strain, e_r, and the associated stresses which may lead to cracking. However, temperature measurements are easy to obtain and to interpret, while strain measurements are much more complex in both respects. As the acceptable temperature limits are used to imply limits on strain, they should, therefore, be variable according to the assumed coefficient of thermal expansion of the concrete, α_c and the restraint to thermal movement, R. The relationship between the factors is demonstrated in the simple equation for evaluating crack risk proposed by Bamforth[8]:

$$e_r = K\alpha_c \Delta TR \qquad (3)$$

where
ε_{tsc} = strain capacity under short-term loading
α_c = coefficient of thermal expansion of the concrete
ΔT = temperature change
R = restraint factor (0 = unrestrained; 1 = full restraint).
K = modification factor of 0.8 which takes account of sustained loading and creep.

For no cracking $e_r < \varepsilon_{tsc}$

Clearly the allowable value of ΔT is inversely related to both α_c and R.

This approach, based on limiting the restrained strain, has also been adopted in CIRIA Report 91, which assumes a value of restraint of 1.0 at the interfaces between new and old concrete, and a modification factor of 0.5. This is consistent with BS 8007[9] for water-retaining structures, which assumes a restraint factor of '0.5 for immature concrete with rigid end restraints, after allowing for the internal creep of the concrete'.

Values of α_c may vary from as low as 7×10^{-6} m/m°C for some lightweight concrete mixes to higher than 12×10^{-6} m/m°C for concretes using siliceous gravel aggregate. Furthermore, the aggregate also has an effect on the strain capacity, ε_{tsc}, (or resistance to cracking) of the concrete, with high values of ε_{tsc} being associated with lower values of α_c. Table 2, from Concrete Society Digest No. 2[3], gives typical values of α_c and ε_{tsc} for concretes using different aggregate types, together with limiting values for temperature drop and temperature differential. **It should be noted that these values are for guidance only**. In practice values of α_c may vary for a particular generic type of aggregate and, where the value is critical, tests should be carried out on a representative concrete mix.

The commonly used value of 20°C as a maximum temperature differential ΔT_{max}, applies to gravel aggregate mixes which exhibit high α_c and low ε_{tsc} relative to concretes using other aggregate types. With the use of a limestone aggregate, for example, which may yield concrete with an α_c as low as 8×10^{-6} m/m°C, higher values of maximum temperature differential may be acceptable. In specifying, ΔT_{max}, therefore, the assumed value of α_c should also be stated, hence defining the limit on differential strain used in the calculation of crack widths and providing a basis for accommodating the use of alternative aggregates.

The values in Table 2 below are for guidance only. Thus where data are available for a particular mix the limiting temperature change can be calculated using the equation:

$$\Delta T = \frac{\varepsilon_{tsc}}{K\alpha_c R} \tag{4}$$

The limiting temperature differential may be derived using the above equation with an assumed restraint factor of 0.36[8].

Table 2 *Limiting temperature changes and differentials to avoid cracking, based on assumed typical values of α_c and ε_{tsc} as affected by aggregate type*

Aggregate type	Gravel	Granite	Limestone	Lightweight
Thermal Expansion Coefficient $\times 10^{-6}$ m/m°C	12.0	10.0	8.0	7.0
Tensile Strain Capacity $\times 10^{-6}$ m/m	70	80	90	110
Limiting temperature change in °C for different restraint factors:				
1.0	7	10	16	20
0.75	10	13	19	26
0.50	15	20	32	39
0.25	29	40	64	78
Limiting temperature Differential °C	20	28	39	55

Restraint may also vary significantly and thus the designer must make some assumptions in his calculations which reflect likely restraints during construction. These will be affected by the selected pour sizes (length and depth), the time between adjacent casts and the sequence of construction.

Guidance on restraint factors is given in CIRIA Report 91, together with a method for designing crack control steel. However, this generally assumes a restraint factor at the joint between new and old concrete of 1.0. No account is taken of the inherent stiffness of the new pour in relation to its immediate surroundings except in the modification factor K, which also takes account of creep and sustained loading effects. ACI Report 207.2R-73[10] gives a more detailed approach for estimating restraint factors in relation to the length/height ratio of a pour, as shown in Figure 1. The restraint at any point is determined by multiplying the restraint at the joints, calculated using equation (5), by the relative restraint at the appropriate proportional distance from the joint, obtained from Figure 1. A comparison between the measured restraint through the height of a bridge pier cast on to a strip footing and values predicted using the ACI method is shown in Figure 2[11], indicating that, provided the assumptions about the relative stiffness of the old and new concrete are appropriate, the method is reasonably accurate. Based on limited measured values of the early age modulus of heat cycled concrete, and the estimated time for the new element to cool, the ratio of $E_n:E_o$ is likely to be in the range 0.7 to 0.8[8] as cooldown occurs.

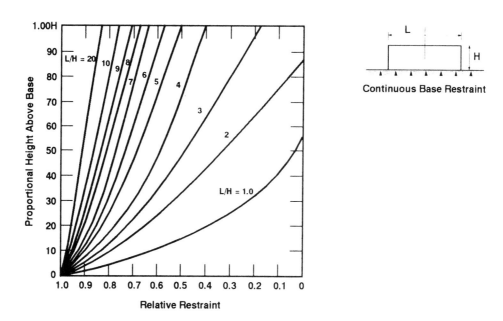

Figure 1 *Restraint factors for elements with continuous base restraint*[10]

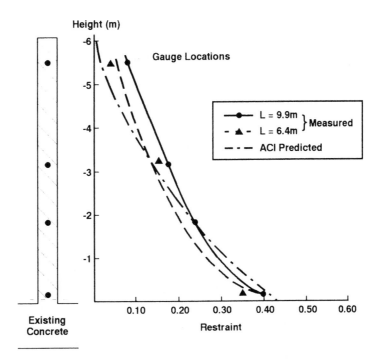

Figure 2 *Measured and predicted restraint in a thick wall cast onto a rigid foundation*

The results in Figure 2 were obtained on the centreline of an 800 mm thick, 6.2 m high, 12 m long bridge pier, cast onto a 1 m deep by 2.85 m wide footing. Hence the restraint at the joint can be calculated as follows:

$$\frac{1}{1 + \dfrac{A_n}{A_o}\dfrac{E_n}{E_o}} = \frac{1}{1 + \dfrac{4.96}{2.85} \times \dfrac{0.8}{1}} = 0.42$$

The gauges were located at proportional heights of 0.3, 0.5 and 0.9, with relative restraints, derived from Figure 1 of 0.53, 0.34 and zero. Hence, predicted restraints through the height, shown in Figure 2, are as follows:

Base 0.42 × 1.00 = 0.42
0.30 0.42 × 0.53 = 0.22
0.50 0.42 × 0.35 = 0.14
0.90 0.42 × 0.00 = 0

The reducing restraint towards the top free surface indicates that the percentage of steel may be reduced with height for early age thermal crack control purposes.

$$\text{Restraint at the joint} = \frac{1}{1 + \dfrac{A_n}{A_o}\dfrac{E_n}{E_o}} \tag{5}$$

where A_n =   cross sectional area of new pour
         A_o =   cross sectional area. of old concrete
         E_n =   modulus of elasticity of the new pour concrete
         E_o =   modulus of elasticity of the old concrete

In some cases, for example, when a high wall is cast onto an existing slab, the designer will be required to exercise judgement with regard to the effective cross sectional areas of new and old concrete used in the calculation. The following rules of thumb may be applied.

- When a wall is cast at the edge of a slab, the relative effective areas may be assumed to be in proportion to the relative thicknesses of the wall and the slab.
- When a wall is cast remote from the edge of a slab, the relative areas may be assumed to be in proportion to the ratio of the wall thickness to twice the slab thickness.

More complex geometries may require more detailed analysis.
Therefore, the designer should define within the specification the following assumptions.

- The allowable temperatures in terms of the maximum value and differentials.
- The coefficient of thermal expansion of the concrete.
- The restraint factors at critical locations. (Where these are based on restrictions on pour sizes, this must also be stated).
- The tensile strain capacity of the concrete.
- Acceptable crack widths, as measured at the surface.

The designer must also consider what actions should be taken in the event of:

1. Unacceptable cracking which occurs within the allowable temperature limits
2. Non-conformance with the temperature limits, but cracking within specified limits
3. Non-conformance with the temperature limits and excessive cracking.

As design codes tend to be conservative, scenario 1 is unlikely, and scenario 3 is clearly the responsibility of the contractor. When scenario 2 occurs, this simply demonstrates conservatism in the design assumption and as experience is gained on a contract, the limits could be adjusted to reflect this.

It is becoming increasingly common, on large civil engineering structures, to specify full-scale trials to obtain performance data on the concrete which can be used to define limits on temperature differentials for use in construction. Where such trials are carried out, care must be taken to ensure that restraints are realistic, particularly in respect of walls cast onto rigid foundation, or slabs which link stiffer elements.

Complex computer models are also available which enable scoping studies to be carried out to investigate the effects of mix type, pour geometry and environmental conditions[12,13] and these are occasionally used for critical structures or elements. However, the value of the output is often limited in absolute terms by the assumptions which have to be made regarding the early age properties of the concrete and their relationship with the temperature history or maturity of the concrete. Validation is also difficult without in-situ measurements of temperature, strain and stress, but trials may often have significant programme implications. This is an area which could benefit from further research.

2.5 PERFORMANCE OF CONSTRUCTION JOINTS

Many large volume pours and deep lifts will be cast as part of much larger elements and this will necessitate the inclusion of construction joints. CIRIA Report 91 – *Early-age thermal crack control in concrete* states that '*with full steel continuity, and assuming that the current cracking control theory for thermal cracking is correct, the crack widths at the joint will be no wider than elsewhere in the section*'.

Hence, if the approach to cracking is to control the crack width and spacing by the design of reinforcement, the location of construction joints is unlikely to be critical. When other means are being employed to limit cracking, the performance of the joint can be assessed by considering the joint to represent the centre line of a pour which is double the length of the pour being cast.

2.6 MATERIALS SPECIFICATION

The designer is responsible for specifying materials that are appropriate for producing a safe, durable structure fit for its intended purpose.

This is usually achieved by placing limits on the properties of the constituent materials and certain mix proportions (see Section 4). In most cases, this will be accomplished by reference to Standards and Codes of Practice. However, the designer should avoid being over restrictive in the choice of materials or limits on mix proportions as, in some cases, this can lead to difficulties with incompatibility between the specified requirements and the practicalities of construction. An example would be specification of an unnecessarily high minimum cement content (on durability grounds) together with stringent limits on the allowable heat build-up within a large volume pour. Consideration of the probable construction techniques should also be an important factor in specification.

2.7 TESTING FOR CONFORMANCE

As with all concreting, conformance testing can be applied both to the constituent materials and the properties of the concrete itself, but in both cases the following factors should be considered.

Sampling – The specification should cover both the methods of sampling and the frequency of sampling. This will depend on the volume of material used on the project and the consequences of non-compliance. Obtaining a representative sample is an essential requirement of any sampling plan.

Testing – The testing requirements should be clearly specified in terms of both what tests are to be carried out (i.e.Standard Test Methods) and how the test results are to be interpreted. The specified tests should be appropriate to the properties being measured and also be capable of being completed in a realistic timescale when considering the rate at which construction will proceed and the consequences of any non-conformance.

Testing will usually cover key properties of the constituent materials (i.e. cement composition, strength and fineness, or aggregate parameters) and fresh concrete properties (i.e. slump, air content or initial mix temperature) as well as the properties of the hardened concrete (i.e. strength).

In many cases, acceptance testing as described above will be undertaken by the contractor. However, conformity may also be achieved by using a concrete supplier who holds current product certification based on product testing and third party surveillance, and who operates an approved quality assurance system to BS 5750. This concept is embodied in the designated mix specification of BS 5328 and is included in draft European Standards. The need for, and extent of, acceptance testing should be agreed between all parties (designer, concrete producer and contractor) prior to construction.

On large contracts or for critical structures, it is worth considering specifying the construction of full-scale trial mock-ups of representative sections, both to assess the effectiveness of the proposed concrete mix and placing techniques and to enable direct measurements of the in-situ concrete properties. In the event of inadequate performance, changes in concrete mix design or construction techniques would be required.

Inspection - Inspection can be divided into three distinct phases, i.e, pre, during and post concrete placing. Where deep lifts or large volume pours are to be cast, the designer should specify that the contractor's method statement makes provision for measurement or inspection as follows:

Preplacing

- Section dimensions
- Reinforcement location (especially cover depth) and bar sizes
- Formwork treatment (adequate and appropriate release agent applied)
- Location of boxouts etc
- Location of embedded inserts
- Location of embedded thermocouples or strain gauges
- Cleanliness

During placing

- Concrete workability
- Concrete temperature
- Rate of concreting
- Compaction
- Open time of live faces
- Effective curing

Post placing

- Standard of finish (i.e. blowholes or sand runs on deep lifts)
- Monitoring of temperatures/strains
- Examination of cracking (both crack spacing and crack widths)
- Achievement of cover

The specification should include both the type and frequency of inspection and the method of analysis and interpretation.

2.8 SPECIFICATION FOR MONITORING OF TEMPERATURES

In large volume pours, the avoidance of excessive temperature rises and differentials is essential to minimise problems of thermal cracking.

The specification of temperature limits is important in this respect, the following being considered:

- Maximum placing temperatures
- Maximum temperature achieved within the section
- Maximum temperature differential between the centre and surface of the section
- Maximum temperature difference between the centre of the pour and ambient temperature
- Maximum temperature differential between successive pours or lifts.

The quantitative limits placed on these parameters should reflect the consequences of any cracking and the properties of the concrete actually used during construction. For example, the type of aggregate used in the concrete has a significant effect on the temperature differential that can be sustained without cracking (some commonly used values are given in Section 2.4) and the effects of peak temperatures on in-situ concrete properties are much less deleterious when composite cements are used.

If temperature limits are imposed, monitoring requirements are normally included in the specification. Embedded thermocouples are an appropriate technique and may usefully be combined with embedded strain gauges (e.g. vibrating wire gauges). The minimum number of thermocouples together with appropriate locations should be specified. Normally this would consist of a thermocouple at the centre of the pour together with additional thermocouples close to the top, bottom and side surfaces of the section. In view of the application, the thermocouples need only be accurate to the order of $\pm 1°C$ and reference calibration is not necessary. It is useful to specify detailed monitoring of the first few pours, relaxing the requirements once a pattern of behaviour has been established (perhaps just monitoring a smaller number of selected locations).

2.9 ACTION IN EVENT OF NON-CONFORMANCE

- Clear details of the action to be taken in the event of any non-conformance with the specification are essential. These should include:
- Test plan for NDT survey and retrieval and testing of cores from suspect areas
- Assessment of consequences of non-conformance
- Action to be taken
  - qualified acceptance
  - local repairs (e.g. sealing, waterproofing, crack injection)
  - removal of suspect concrete
  - additional strengthening measures.

3 Planning

This section provides a detailed check list for the contractor at the planning stage of construction using deep lifts and large volume pours.

The successful completion of a deep lift or a large volume pour depends on co-operation between the designer, the contractor and the concrete producer. In addition, it is essential that planning begins at an early stage in order to forestall potential problems. For large volume pours the following points need to be resolved at least two to three weeks before the pour is to take place:

- Establish authority, responsibility, planned activities and communications for each party concerned
- Agreement on approved material sources and mix details
- Provision for standby equipment
- Access to the site
- Setting a 'fixed' start date

Where deep lifts also involve large volumes, all of the above still apply. However, the essential differences between deep lifts and large volume pours are:

- The much higher rate of casting normally associated with the large volume pour
- The restricted access for concrete delivery and compaction in the deep lift.

The most important point is to ensure that there is adequate back-up plant available to maintain continuity in all stages of delivery, distribution, compaction and finishing of the concrete.

3.1 CAPABILITY

Although every site is unique, and every pour will have its own special problems, previous experience in planning and undertaking the placing of concrete in deep lifts or large volume pours is invaluable. The key to success is achieving continuity of casting at a rate which avoids cold joints and other construction defects, using methods which are consistent with recognised good practice. Hence, all those involved in the casting must be made aware of the importance of their link in the chain of:

- production
- delivery
- distribution
- compaction
- curing

In the case of a large volume pour, other trades, e.g. steel fixers and carpenters, must also be made aware of the 'fixed' start time, as substantial plant and manpower has to be committed to the pour itself, much of which may be sub-contracted.

The contractor can demonstrate his capability by providing a work procedure, defining management, method, programme and inspection and testing regime, which takes account of the unique features of the pour to be cast.

3.2 PLANNING MEETING

For the benefit of those on a team who have not previously experienced continuous concreting beyond normal working hours, or the special procedures involved in deep lifts and large pours, there should be formal meetings between all involved at all levels. These should inform and advise, the objective being to highlight critical aspects of the construction process, and in so doing to forestall potential problems.

3.3 PLANT

For **deep lifts**, the main considerations with regard to plant are:

- the limited space on the working platform at the top of the lift
- the limited access for concrete distribution and compaction and possibly poor visibility within the pour
- the high pressure on the formwork, and the need to prevent leakage.

As placing rates are unlikely to be excessive, the delivery of concrete to the pour may be by any of the conventional methods, e.g. crane and skip, pumped. However, boom pumps have an advantage in that the line takes up less space and requires less handling.

To distribute concrete to the bottom of the form without segregation, tremie pipes, drop chutes or flexible hosing are commonly used. In the latter case, this may be fixed to the end of the pump boom.

Vibrators will require longer lines than normal, and different sized pokers may be needed if there are areas of congestion. As it is difficult to control the location of a poker at the bottom of a deep lift, it may be appropriate to consider jigs with fixed vibrators which are raised as the lift progresses. Because of the difficulty of raising pokers from the base of a deep lift before inserting them in a new position, a greater number of pokers than normal may be required.

External vibrators may also be used to overcome access problems. These may either be attached to the formwork in fixed positions, or on demountable fixings which would allow a smaller number of vibrators to be moved up the formwork as the height of the concrete rises. It should be noted, however, that formwork should be specifically designed with external vibration in mind and should be both rigid enough to withstand the effects of the vibration and capable of transmitting this vibration to the concrete.

Formwork design is also important with regard to stability, stiffness and the avoidance of grout leakage at joints.

Lighting may also be required in very deep lifts where visibility is poor.

In **large volume construction** the main considerations are:

- maintenance of a high rate of placement
- distribution over a large area

As indicated in Table 3, most large volume pours rely on pumps to achieve high rates of concrete placement[6,14-18]. In Germany in 1988, a record 17 000 m^3 pour was cast continuously using four pumps over a period of 78 hours[6]. The average placing rate per pump was about 55 m^3/hr, and 90 ready mix trucks were used, operating from six batching plants. Higher placing rates of 100 m^3/hr per pump were achieved in the US in 1990, where a 5700 m^3 pour was cast in only 11½ hours[14].

The number of pumps needed will depend not only on the placing rate to be achieved, but also the plan area of the pour. Mobile pumps have folding distribution booms which vary in length from 15 m to over 50 m[18]. Hence boom pumps are particularly suitable for large area, large volume pours.

Table 3 *Some examples of large volume pours*

Ref No.	Location/ Date	Pour details		Time taken (hrs)	Average rate of placing (m³/hr)	Plant used	Average rate per pump (m³/hr)	Concrete mix details
		Volume (m³)	Thickness (m)					
6	Frankfurt, Germany 1988	17000	8.5	78	218	4 pumps	55	Not given
14	California, USA 1990	5700	1.7 to 3.8	11½	496	5 pumps	99	Grade 20 pfa
15	Seattle, USA 1989	8230	4.7	13½	610	9 pumps	68	Grade 40, 30% pfa
16	Jamshedpur, India 1990	3600	4.3	44	82	6 pumps	15	Grade 25, BFS Cement, W/C = 0.42
17	Thames Barrier 1982	6600	5.0	72	92	Tremie	-	Grade 30 50% pfa
18	Sheffield 1976	3000	2.0	22	136	2 pumps	68	Not given

Typically, pumps can achieve practical placing rates of between 30 and 100 m3/hour, compared with average rates of the order of 8-12 m3/hour with crane and skip[19] and are, therefore, much more suited to large volume concrete construction.

Other placing methods may be appropriate when the plan area is smaller. For example, chutes provide a rapid method of distribution over relatively short distances and are most appropriate when high workability concrete is used. Conveyors are more suitable for use with low workability mixes.

When night working is expected, daylight standard lighting must be provided, particularly for reasons of safety.

In view of the importance of maintaining continuity of placement, back-up must be available for all essential items of plant.

3.4 LABOUR

Sufficient labour must be allocated to each stage of the casting process so that construction proceeds without interruption.

Lengths of shifts and changeover times must be agreed and, on long shifts, back-up teams must be available to accommodate work breaks.

It is preferable that experienced teams are used, and additional training must be given where appropriate.

3.5 MATERIAL SUPPLY

As continuity is an essential feature of the successful construction of deep lifts and large pours, the benefits of advanced planning in relation to all materials suppliers cannot be overstated.

If ready mix concrete is to be used, the supplier, who will also have to meet the needs of other customers, must be given sufficient advanced warning when high placing rates are required. When the casting rate necessitates the use of more than one batching plant, this requires a high level of coordination by the concrete supplier. However, even when the casting rate is well within the capacity of a single plant, it is still advisable to utilise more than one, to ensure continuity in the event of one plant being unable to maintain supply.

When very large volumes are to be placed, or very high placing rates are required to meet the programme, it may not be appropriate to rely solely on site production, except on very large sites which have more than one batching and mixing plant. However, site batching plant can be used in conjunction with a ready mix supply. In this case, the contractor must advise all material suppliers well in advance and obtain assurances that the necessary rates of supply will be met.

3.6 ACCESS

Access is important in a number of ways.

- For the concrete supplier to the site
- Within the site to the pour
- To the point of delivery within the pour.

Access to the site will depend not only on geography, but also the time of day, day of the week, and possibly the time of year. Consideration must be given at the planning stage to possible delays due to heavy traffic, road works, etc and the benefits which may be achieved by working non-standard hours. For example, in some locations casting during the night or at weekends may allow easier access while in other cases the opposite may be true.

On site, a temporary parking area is required for the ready mix trucks, routes must be carefully coordinated, particularly if there are several delivery points, and an area must be allocated for the trucks to wash out.

Access to the point of delivery within the pour is determined by the selection of plant, and has been discussed in Section 3.3.

3.7 TIMING

The timing of the pour may be significant in a number of ways.

- It will determine ease of access of the site
- It will determine availability of plant and concrete supply from sub-contractors with many other customers
- It will have an effect on the workability of the concrete and subsequent performance in-situ.

The importance of appropriate timing in relation to access to the site is covered in Section 3.6.

Availability of plant will also be determined by the timing of the pour. For example, at night and weekends there is less likely to be a high demand on concrete supply and it may be possible, therefore, to dedicate particular plants. This also helps to achieve consistency of supply.

Two recent examples of very large volume pours in the US, cast at high placing rates over relatively short periods – 5700 m^3 in 11½ hours[14] and 8230 m^3 in 13½ hours[15] – involved night-time starts at 10.00 pm and midnight respectively.

Technical advantages can also be gained by night-time concreting associated with lower ambient temperatures and the lower concrete mix temperatures than those occurring during the day. Particular benefits include:

- slower rate of workability loss of the fresh concrete
- longer stiffening time, reducing the risk of cold joints
- slower rate of generation of heat of hydration, from a lower start point, thus reducing the risk of early age thermal cracking.

3.8 QUALITY ASSURANCE

Quality assurance systems are designed to prevent mistakes occurring during construction and also to identify and remedy any problems and non-conformances that do occur. They are, therefore, of great benefit in ensuring successful construction. Such systems (preferably complying with the requirements of BS 5750) should encompass all measures necessary to maintain and regulate the quality of the construction process.

A quality system should include inspection and testing of materials and fresh and hardened concrete, as well as documented procedures for production, placing and curing of concrete.

The testing plan should also form part of the quality assurance system. To be effective, quality assurance should involve designer, contractor and materials (or product) suppliers.

Where suppliers operate a third party certificated quality assurance system, the contractor should assure himself that this system adequately covers all the operations for the particular project in question. Only if this is the case should acceptance testing of the project be waived.

Close cooperation between all parties involved is a prerequisite of operating an overall quality system and for successful construction as a whole.

3.9 MONITORING

It should be clearly stated in the specification where special monitoring, in addition to the normal inspection by the engineer prior to casting, and control of the concrete during casting, are required. The most common 'additional' requirements for thick walls or large volume pours, are for temperature measurements. In such cases, the following should be agreed:

- The precise locations of the measuring points in relation to the limits given in the specification (see Section 2.4)
- The protection of the measuring points prior to and during concreting
- Calibration standards and accuracy
- The frequency of measurement, and for how long measurement must continue
- The form in which the results are to be presented
- The way in which the results are to be interpreted
- Action limits, and the procedures to be initiated when these limits are reached
- Actions to be taken in the event of non-compliance with the specified procedures or non-conformance with specified performance limits.

3.10 CONTINGENCY PLAN FOR LOSS OF CONCRETE SUPPLY

Although unlikely, there is always a finite risk of loss of materials supply, however well the project is planned and however much back-up is provided. The contractor is therefore advised to have a contingency plan for stopping the pour at any stage and incorporating joints.

The contingency plan should include:

- preferred locations of joints, with advice from the designer
- the method for forming joints

This plan should be prepared in conjunction with the planning of the concreting sequence.

3.11 CONTRACTOR'S CHECK LIST

In the planning stage, the contractor should pay particular attention to the following:

1. Advanced warning to all parties involved, and in particular, the material supplier. The local community, police etc should also be notified.

2. Concrete production – if ready mixed concrete is to be used the contractor must check:
    - Location of the plant in relation to transit time and risk of delays (e.g. in rush hours)
    - Throughput as determined by storage capacity, mixing rate, number of delivery trucks

- Quality control in relation to risk of rejecting non-compliant mixes
- Back-up plants, or use of several plants.

3. Access to the site – separating incoming and outgoing mixer trucks, provision of on site space for queuing and washdown.

4. Concrete distribution – appropriate selection of plant to achieve the required placing rate. Pumps are most commonly used for very large volume placements, but skips, chutes and conveyors are also used. Arrangement of back-up plant in case of breakdown. Lighting may be needed for deep lifts and is essential for night-time concreting.

5. Concrete placing sequence – planning of sequence of placing from one or more points, and thickness of layers.

6. Placing and compaction – ensure sufficient labour and plant for thorough compaction.

7. Finishing – this may be delayed if the mix has a slow set.

8. Condition for non-standard working hours – check local by-laws for night and weekend working. Agree conditions with labour force, e.g. shift working, cover during breaks.

9. Quality control – rate of sampling of concrete; storage of test specimens; actions in the event of non-conformance. In the case of concrete subject to third party certification, the quality assurance scheme of the concrete producer must be examined and approved by both contractor and designer.

10. Temperature monitoring – location of measuring points; method of recording temperature; agreed action limits; actions in the event of non-conformance.

11. Achievement of specified temperature limits – mix selection, trials to measure temperature rise or heat of hydration, cooling of the fresh concrete, in-situ cooling with embedded pipes, thermal curing.

12. Contingency plans for:
    - delays or failure of concrete delivery
    - breakdown of plant for concrete distribution and compaction
    - extension beyond anticipated pour time (particularly if running beyond normal working hours)
    - severe weather, especially heavy rain.

4 Concrete mix design

This section provides guidance on the selection of concreting materials and mix design to achieve performance characteristics required for the successful completion of deep lifts or large volume pours.

Good concrete mix design is an essential part of any successful concrete construction project, but placing concrete in deep lifts and large volume pours requires certain additional considerations. The final concrete mix design must be capable of satisfying both the requirements of the designer in terms of structural and durability properties, and the needs of the contractor for production and placing the concrete in deep lifts and large volume pours.

4.1 RESPONSIBILITIES

The responsibilities for ensuring that the concrete is capable of producing a safe durable structure, fit for its intended purpose, is shared between the designer, the contractor and the concrete producer.

The designer is responsible for specifying those properties that are required to ensure the safe structural performance and durability of the structure. These are typically limited to the characteristic strength, durability parameters such as a maximum water/cement ratio or minimum cement content, and cement type (often by reference to standards or Codes of Practice), and limitations on maximum temperatures or temperature differentials as a means of limiting crack widths.

As already highlighted in Section 2.1, it is important for the designer to recognise that an over conservative limitation on minimum cement content, which is generally perceived to increase durability, may actually have the opposite effect. This occurs in cases where heat of hydration is likely to lead to problems such as early age thermal cracking or excessively high temperature build-up on the surface.

There is no evidence that increased cement contents, per se, as opposed to their effect on reducing w/c ratio, are of benefit for durability. Within reason, therefore, minimum cement content should not be specified for durability requirements.

In general the concrete specification, in accordance with BS 5328[21], will fall into one of the following categories:

- designed mix
- prescribed mix
- standard mix
- designated mix.

Considering the number of factors that need to be taken into account for successful construction in deep lifts or large volume pours, it is only really appropriate to specify designed mixes.

The contractor also has specific requirements for distribution and compaction of concrete. These may not always be compatible with other requirements, however. For example, a low cement content is desirable for low temperature rise and low risk of cracking, but if the concrete is to be pumped, the cement content should normally be not less than about 290 kg/m^3[20].

The designer must, therefore, be aware of possible constraints on mix design when specifying an allowable maximum temperature. As a maximum value of 70°C is often specified, this problem is unlikely to arise except in extreme conditions of summer concreting with a very rich mix. Nevertheless, the designer must be aware of the possible constraints which may arise when designing for the use of high grade concretes in thick sections.

Furthermore, the designer must also state what assumptions have been made in the design process. These will include:

- the materials safety factor applied to the design strength
- the aggregate type as it affects the thermal expansion coefficient and tensile strain capacity of the concrete, and hence the risk of early age thermal cracking
- the temperature drop used in the calculation of crack widths
- restraints, as affected by pour sizes and construction sequence.

As some of these assumptions may differ from practice once a contract is underway, it is essential that they are stated in the contract documents. This will allow a sound technical dialogue between the designer and the contractor when they discussing the consequences of changes, and consider the suitability of mixes in relation to the particular construction programme.

The detailed mix design will be produced by the contractor or concrete producer and will combine the specifier's needs with the practical requirements for successful placing of the concrete.

Test results confirming the properties of the proposed mix (usually limited to compressive strength and workability) are often submitted for approval together with the mix proportions.

4.2 PROPERTIES TO BE CONSIDERED

When designing a concrete mix for use in deep lifts or large volume pours, the following properties should be considered:

1. **Specified properties**

- Characteristic compressive cube strength
- Minimum/maximum cement content
- Durability requirements – resistance to sulphates, chlorides, freeze/thaw, ASR
- Limiting temperatures and temperature differentials.

2. **Construction properties**

- Workability and workability retention – to enable the concrete to be transported, placed and compacted. This must take account of the selected method of distribution, e.g. pump, skip, tremie
- Stiffening time – as it affects formwork pressures and cold joints
- Cohesiveness and bleed – particularly important in deep lift construction
- Strength development – as it affects formwork striking times
- Heat evolution – in relation both to the risk of cracking in large volume pours and to the resulting in-situ properties
- Finishing characteristics.

Some of these properties are not covered by standard test methods, but quantitative assessments can be derived from site trials.

The general principles of mix design are covered elsewhere[22,23].

4.3 FACTORS INFLUENCING KEY PROPERTIES

Given that all the specified properties of the concrete mix have been achieved, construction in deep lifts and large volume pours involves the following factors.

4.3.1 Section dimensions

The section dimensions influence the mix design in a number of ways, the minimum dimension affecting both the selection of maximum aggregate size and cement type and content.

To avoid problems with aggregate arching, BS8110[24] recommends that the nominal maximum size of coarse aggregate should be not greater than one quarter of the minimum section thickness. The spacing of the reinforcement must also be taken into consideration and it is generally accepted that the maximum aggregate size should be less than three quarters of the minimum clear spacing between reinforcing bars, bundles of bars or prestressing strands.

The minimum section thickness also governs the rate at which heat generated by cement hydration can be dissipated. Some specific figures for use in design are given in CIRIA Report 91[7]. Thicker sections have the potential for considerable internal temperature rises (see Section 6) and the use of concrete with either lower total cement contents or composite cements may be appropriate[3]. Narrow sections with congested reinforcement require concrete to have a high workability if placing without honeycombing or excess voids is to be achieved.

Concreting deep sections often involves placing concrete from the top of the form, and in order to minimise segregation as the concrete falls through the formwork and reinforcement cage, the mix must be cohesive. This influences the proportions of fine to coarse aggregate as well as the consistency of the cement paste.

The effects of bleed and settlement are of greater significance in very deep sections and again point to the need to achieve a cohesive concrete mix.

4.3.2 Access and placing techniques

The technique proposed for placing the concrete and the ease of access to the structure can both influence the requirements of the mix design. When access is unrestricted, concrete may be placed by skip or directly from a readymix truck. In more restricted situations, pumping or tremie placement is often preferred. The transport time between mixing and placing will also influence the detailed mix design.

Concrete placed by skip, or by chute directly from the readymix truck frequently needs no additional considerations over and above the usual requirements for adequate workability. Successful pumping or tremie placing, however, both require cohesive concrete of medium/high workability if blockages are to be avoided. For pumping, a minimum fines content (cement + fine sand) of 400 kg/m^3 and slump of 75-100 mm is often recommended[20]. For concrete to be placed by tremie a slump of 150 mm is commonly used.

If concrete is to be transported or pumped over long distances, retention of workability must also be considered. A reduction in workability may be due to loss of water by evaporation, in which case it may be acceptable to add water prior to discharge. Where long haulage times are expected, an acceptable method of adding small quantities of water, as permitted by BS 5328: Part 3 to restore workability, should be agreed in advance. Alternatively, retarding admixtures and/or the selection of cements with longer setting times, e.g. composite cements with pulverised fuel ash (pfa) or ground granulated blastfurnace slag (ggbs), may be used.

4.3.3 Volume of pour and rate of placing

The volume of pour in itself does not influence the mix design, although the minimum dimension will, as discussed in Section 4.3.1. However, the rate at which concrete is to be placed will control the acceptable workability loss between batching, arrival at site and placing. For information on concrete properties in relation to transportation and handling, reference should be made to CIRIA Funders Report CP/18[25].

The stiffening time of the concrete must be compatible with the rate of placing. If the concrete sets too rapidly (relative to the placing rate), in addition to the formation of cold joints, there is also a risk that vibration may damage the lower levels of the lift. In such cases, consideration should be given to the use of slow stiffening concretes, i.e. concretes containing composite cements[26] or the incorporation of retarding admixtures[27]. This will be of most significance in large volume pours.

However, for deep lifts the effects of set retardation on formwork pressures are likely to be of greater significance. This is addressed in CIRIA Report 108[28].

4.4 CONCRETE MIX DESIGN FOR DEEP LIFTS

As can be seen from the discussion above, certain criteria for concrete mixes for successful placing in deep lifts must be met. The particular properties to be considered are:

- workability
- cohesiveness and freedom from bleed
- stiffening time

4.4.1 Workability

Workability is primarily controlled by the water content of the mix. Hence, for mixes without admixtures, higher workability requires more water and thus more cement to achieve a specified strength grade.

Concrete that is to be placed in deep lifts must be workable in order that it can be properly compacted in narrow forms and around congested reinforcement. This is important as the difficulties of compacting concrete at the base of deep forms may mean that internal vibrators have to be inserted at wider spacings than normal.

The introduction of water reducing admixtures (plasticisers and superplasticisers), has enabled high slump concretes to be produced at high strengths and without the need for excessive cement contents[28]. Workability is also improved by selecting the maximum aggregate size compatible with the section dimensions and reinforcement spacing and ensuring that both coarse and especially fine aggregates are properly graded[22].

The fines content must also be high enough to avoid harshness. It should be noted that concretes containing pfa to BS 3892: Part 1[29] have improved workability at a given water content, due to both the spherical shape and the grading of the pfa particles and its lower density which results in a higher volume of paste for a given weight of composite cement[26]. Modest improvements[26] can also be achieved using ggbs to BS 6699[30].

It should be noted that when high quantities of admixture are used to produce workable concretes at low w/c ratios, there may be a tendency for the concrete to be very 'sticky' and resistant to flow under the influence of vibration even though the slump is suitably high. Excessive levels of plasticising admixtures may also lead to segregation and retardation.

4.4.2 Cohesiveness and bleed

Concrete placed in deep lifts is often poured from the top of the forms, and even when other techniques such as pumping and tremie placing are used, there is always the risk of the concrete segregating as it strikes the formwork and reinforcement. This can cause separation of aggregate from the mortar as the latter adheres to the reinforcement, and consequent honeycombing. Increasing the cohesiveness of the mix will minimise the risk of separation.

Cohesive concretes are characterised by a higher than average proportion of fine aggregate and a high cement content. Addition of pfa to the concrete, which results in a much lower water/composite cement ratio by volume, also improves cohesiveness. The use of fine sands tend to produce more cohesiveness than equivalent quantities of coarse sands, but this would need to be associated with increased admixture doses to maintain workability.

A well proportioned pump mix is often suitable for use in deep lifts. This combines the cohesiveness and workability needed to avoid blockages in the pipeline. A rule of thumb[20] is to add an additional 4% of fine aggregate over the optimum content for a normal mix[22] of given workability.

Closely allied to cohesiveness is reduced bleed and settlement. Bleed is the upward separation of water from the concrete under the influence of gravity. The results of bleed (and associated settlement of the concrete) are a general increase in the water content in the upper parts of a lift, together with water filled voids under reinforcing bars and cracking in areas where settlement is restrained[31] .

Deep lifts are particularly prone to the effects of bleed, as the amount of settlement is generally proportional to the depth of the section.

As a rule of thumb, bleed may be excessive if the aggregates are very clean (i.e. the salt content is low), if less than 20% passes the 300 micron sieve, and if the material tends to be single sized. Hence, bleed can be minimised by ensuring that the aggregates are properly graded and in particular, by including a sufficient quantity of fine material. This may be achieved by increasing the proportion of fine aggregate, by the use of a sand with a high proportion of material finer than 150 micron, by using a more finely ground cement or by incorporation of an increased volume of cementitious materials such as pfa, or by the addition of a much finer material such as microsilica[32]. While pfa retards the set, and the period of bleed is extended, the rate of bleed is usually reduced sufficiently to more than offset this[26]. Microsilica concretes in particular may exhibit very low bleed to the extent that exposed faces are very sensitive to curing and horizontal surfaces are difficult to finish. Concretes containing a high percentage of ggbs are sometimes more prone to bleed[26].

There is an optimum fine aggregate grading and content leading to a minimum void content and reduced bleed (Figure 3) and if this is exceeded, the bleed may actually increase. The optimum sand content can be estimated simply by weighing the combined aggregate in a bucket of fixed volume. As the sand content is varied, the weight (or bulk density) will also change and the maximum weight indicates the minimum voids. These aggregate proportions would then be used in subsequent trial mixes.

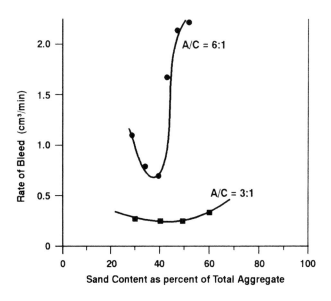

Figure 3 *The influence of sand content on the rate of bleed (TEL Data)*

A reduction in the water content of the mix will also reduce bleed, but where this is achieved with admixtures which also retard the set, this may partially offset the benefit by increasing the time over which bleed can occur[33],as shown in Figure 4.

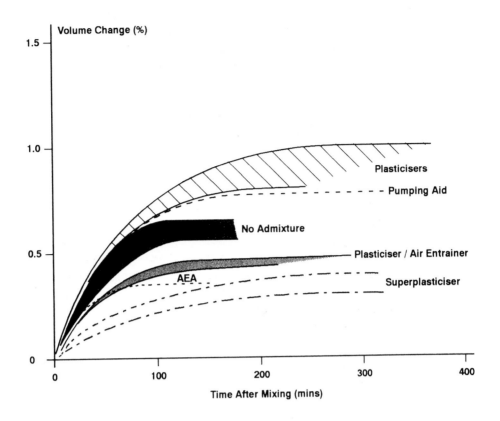

Figure 4 *The influence of chemical admixtures on the settlement of concrete (TEL Data)*

Another effective remedy for reducing bleed is air entrainment. The entrained air bubbles act in a similar way to very fine aggregate particles thus decreasing bleed, but without a deleterious effect on workability. However, air entrainment is not always justifiable, either economically or practically as it causes a reduction in strength, which must be offset by reducing the w/c ratio, and this may require an increase in cement content. It can be valuable, however, if the available materials are such that excessive bleed is unavoidable. The relationship between bleed, air content and w/c ratio is shown in Figure 5.

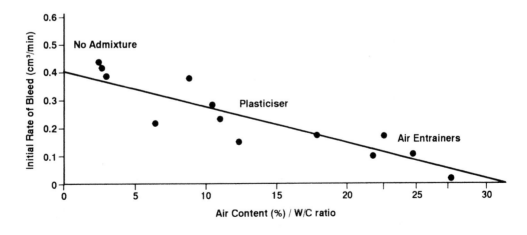

Figure 5 *A typical relationship between the initial rate of bleed and air content and w/c ratio of the concrete (TEL Data)*

In view of the numerous factors which influence bleed and settlement, trial mixes should be undertaken to produce the optimum mix proportions for minimising bleed. Full-scale placing trials prior to construction are also recommended for identifying any potential problems.

4.4.3 Stiffening time

When concrete is placed in deep lifts it is often advisable to avoid rapid stiffening. This enables the upper layers to be compacted without damaging lower layers by transmitted vibration. The risk of cold joints is also reduced. However, excessive retardation of set can lead to increased bleed and increased formwork pressures.

Concretes containing pfa (which are particularly useful in deep lifts due to their resistance to bleed and generally cohesive nature) will exhibit retarded set of roughly half an hour for each 10% of pfa by weight of the composite cement. Ggbs has very little effect at replacement levels below 50%[26], but will exhibit a retardation of about 1½ hrs at the 70% level.

However, admixtures are often the most influential factors affecting concrete stiffening time. Even if retarders are not used, plasticisers based on lignosulphonates often cause retardation and superplasticisers at high levels of addition may act in a similar manner[27].

4.4.4 Early age thermal cracking

Although the occurrance of early age thermal cracking is not confined to deep lifts, many such pours are designed for containment, and cracking must, therefore, be controlled. The problem is of greater concern for large volume pours and aspects of mix design for minimising the risk of cracking are covered in Section 4.5.1.

4.4.5 Summary

To summarise, in addition to adequate strength and durability, an appropriate concrete mix design for successful placing in deep lifts should have the following characteristics:

- Workability appropriate to the method of compaction and degree of reinforcement (probably in excess of 75 mm slump).
- Good cohesiveness produced by a suitably high level of cement and well graded fine aggregate.
- Low bleed by achieving low water/fines ratio or by the use of air entrainment.
- Extended stiffening time (but not excessively retarded) to minimise cold joints.
- Early age thermal characteristics compatible with the acceptable level of cracking.

4.5 CONCRETE MIX DESIGN FOR LARGE VOLUME POURS

In addition to the normal requirements for strength and workability, concrete for placing in a large volume has certain specific requirements regarding thermal characteristics in order to minimise the risk of thermal cracking. It should also be noted that the requirements for cohesiveness and bleed described in Section 4.4.2 may also apply to large volume pours if they are particularly deep, or if placing through the reinforcement is unavoidable and delayed stiffening time is especially important to avoid cold joints. The following additional factors must be considered in relation to the risk of cracking due to early age thermal effects:

- heat evolution
- thermal expansion coefficient
- tensile strain capacity

4.5.1 Heat evolution

Hydration of cement is an exothermic reaction and in large volume pours where heat dissipation is low, the temperature within the pour can rise significantly. In the centre of sections greater than 2 m thick, the temperature rise will be nearly adiabatic and be proportional to the cement content of the concrete mix[34]. Values of maximum temperature of the order of 60-70°C are common.

In smaller pours where heat is more readily lost to the environment, the temperature rise is also affected by the rate at which heat is developed[7]. The rate and amount of heat generated by Portland cements to BS 12 depends on the fineness of grinding and the chemical composition. For example, there are data which indicate that cements with a high C_3A content will develop the highest temperature rises[35,36]. Conversely, cement with a low C_3A content, e.g. sulphate resisting cement, will tend to develop less heat.

With regard to concrete mix design, there are several factors which will influence both the rate of hydration of the cement, and the ultimate heat generated:

- Total content of cementitious material
- Type and source of Portland cement
- The type and proportions of composite cements which utilise either pfa or ggbs.

Other factors, such as the type of formwork, the geometry of the pour and the concrete mix temperature are also influential. These are beyond the scope of this section on mix design, and are discussed later in Sections 5 and 6 appropriate to the nature of the pour being cast.

Cement content

In the centre of a large volume pour, which is close to adiabatic conditions, the temperature rise is approximately proportional to the cement content. Hence a reduction in cement content will effect a proportional reduction in temperature rise (Figure 6).

Water reducing admixtures are beneficial as, for a concrete of given workability, a reduction in cement is possible without loss of strength, i.e. the w/c ratio can remain unchanged. Plasticisers will permit a reduction in water and cement of the order of 6-8% while reductions of up to 20% may be achieved with superplasticisers[27].

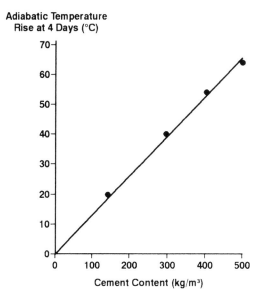

Figure 6 *Adiabatic temperature rise as affected by cement content (TEL Data)*

An increase in the maximum aggregate size compatible with the section dimensions and reinforcement spacing will also contribute to a reduction in total cement content. However, there is evidence that this will also cause a reduction in the strain capacity of the concrete, and that this will more than offset the benefit of reduced temperature rise in relation to risk of cracking[37]. This is illustrated in Figure 7.

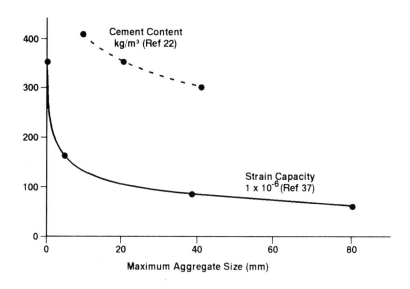

Figure 7 *The influence of aggregate size on the strain capacity and the cement content of concrete*[37]

A reduction in cement content can also be achieved by the use of microsilica, which is particularly effective in lower grade concretes. For strength, the cementing efficiency of microsilica is about four, being part chemical (reactive silica) and part physical (pore blocking of fine particles), while its contribution to heat of hydration is only one to three times that of Portland cement (PC)[32]. Hence lower temperature rise can be achieved by the use of microsilica concrete, compared with PC concrete of the same strength grade.

Composite cements

Composite cements based on combinations of Portland cement with either pfa (Pulverised fuel ash to BS 3892 Part 1) or ggbs (ground granulated blastfurnace slag to BS 6699) generally hydrate at a slower rate than Portland cements to BS 12 and produce a lower temperature rise. This applies to both manufactured blends, and blends produced in the mixer. The level of pfa in such blends is usually up to 40%, whereas ggbs is used in larger amounts, often up to 75% of the blend. The relative effects of different composite cements on the typical temperature rise in sections of different thickness is shown in Figure 8.

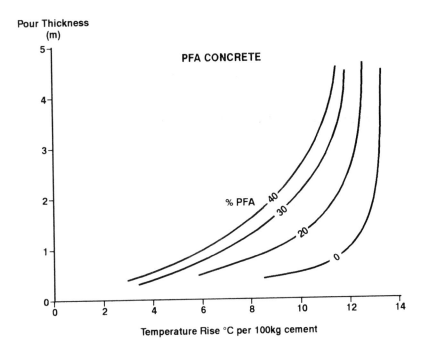

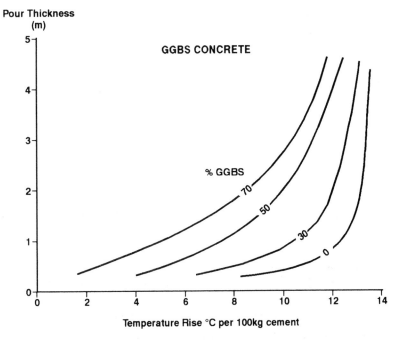

Figure 8   *The influence of pfa and ggbs on the temperature rise in concrete elements of different thickness*[34]

It must be appreciated, however, that not all ggbss exhibit the same performance. Some current sources are more reactive than those commonly used over the last 20 years, and provide less benefit in terms of reduced rate of heat generation[38], as shown in Figure 9. Similarly, blends of the various sources of pfa and PC may result in different hydration characteristics. It is, therefore, advisable to seek the manufacturer's advice, or to obtain test results for the proposed materials and mix designs.

Numerous methods are available for measuring the heat generation or temperature rise in concrete. For thick sections adiabatic or near adiabatic testing is preferred[35,36] as this represents most closely the conditions at the centre of the element. For practical site purposes, it is common to cast an insulated 1 m^3 block and to measure the temperature rise at the centre. A simple plywood form lined with 50 mm thick expanded polystyrene is adequate to provide comparative results.

It must also be appreciated that, while the weight for weight temperature rise of composite cements is reduced, the benefit may be partially offset by the need to use higher total cement contents if the concrete is required to meet the same 28-day standard cured strength.

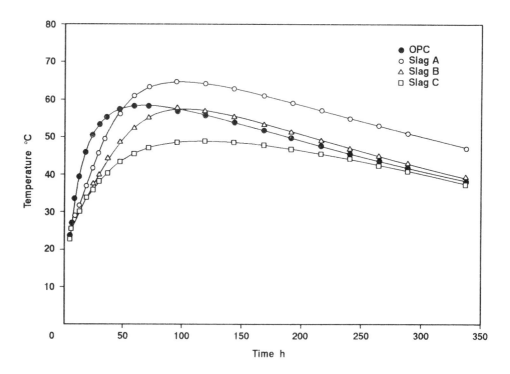

Figure 9 *The effect of different sources of slag on the temperature development in laboratory simulated 3 m deep pours*[38]

Influence of early age temperature on strength

While all concretes are adversely affected by high early age temperatures, there is now substantial evidence that in thick sections, the in-situ (heat cycled) strength of composite cement mixes will be higher than PC concrete designed to the same grade. Results from six studies[34,39,40,41,42,43] are shown in Figure 10 for pfa concretes. In the most extreme cases, the difference in in-situ strength was in excess of 20 MPa, with more typical values of the order of 10 MPa. Additional benefit could be achieved, therefore, by reducing the materials safety factor, γ_m, for concretes containing pfa according to the predicted temperature rise in the element, hence enabling a lower grade of concrete to be used.

Table 4 shows typical mix proportions for concretes designed in accordance with the DOE mix design method[22]. PC mixes have been designed to achieve mean strengths of 30, 40 and 50 MPa and PC/pfa mixes (with 30% pfa) have then been designed with the same total cement content. The standard 28-day strengths of the latter were 5 to 8 MPa lower. This is of the same order as the benefit achieved as a result of the in-situ heat cycle for a temperature rise of about 30°C (i.e., consistent with 300 kgs of PC/pfa in a thick section). This indicates that mixes with the same total cement content using either PC or composite cement with PC/pfa, will achieve the same in-situ strength at 28 days.

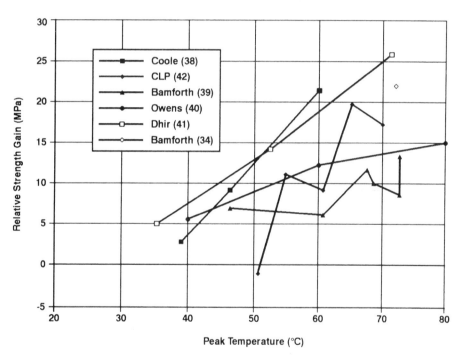

Figure 10 *The relative effect of pfa on the heat cycled strength of concrete, compared with Portland cement concretes*

Thus in sections which will be subjected to a significant rise in temperature due to hydration of cement, it is suggested that the materials safety factor, γ_m, can be reduced to 1.25 when pfa is used in the mix, or alternatively the strength grade may be specified at a later age, say 56 or 90 days.

Similar, but less substantial benefits may be achieved using ggbs[34,38], although at this time insufficient data are available to enable specific recommendations regarding the change in γ_m.

The long-term strength gain of the pfa concrete is greater. Measurements obtained by the Building Research Establishment on cores taken from structures up to 33 years old[44] demonstrated that 'the ratio of core strength to 28-day cube strength ranged from 146% to 240% (mean 193%) for the pfa concrete, and from 116% to 150% (mean 138%) for the PC concrete'.

The lower temperature rise and the enhanced resistance of composite cements to sulphate attack[26] may also reduce the risk of delayed ettringite formation and its disruptive effects.

Table 4 *Typical effect of composite cement (70% PC/30% pfa) on 28-day strength at a fixed total cement content*

PC content (kg/m3)	290	345	410
W/C	0.62	0.52	0.44
Mean cube strength (MPa)	30	40	50
PC/pfa Content (kg/m3)	290	345	410
W/(C+F)	0.57	0.48	0.40
Mean cube strength (MPa)	25	32	43
Reduction in standard cube strength(MPa)	5	8	7
Enhancement under thermal cycling(MPa)	6	8	10
Net benefit of pfa on heat cycled strength (MPa)	1	0	3

The concrete mix design should, therefore, contain the minimum cement content compatible with the specified strength and durability requirements, and preferably use a low heat composite cement. In addition, the designer should consider compliance based on a later age strength, to take benefit from the continuing strength gain of concretes using these cement types.

4.5.2 Thermal expansion coefficient and strain capacity

Thermal contraction cracking is discussed in greater detail in Section 6. However, a reduced thermal expansion (and contraction) coefficient of concrete, when combined with a high tensile strain capacity, is of great benefit in reducing the potential risk of cracking.

The aggregate fraction of concrete (both coarse and fine) makes up the majority of its volume and thus exerts a major influence over the thermal expansion of the concrete. Different aggregate types produce concretes with different thermal expansion coefficients and some typical values[3] are given in Table 2. Limestone aggregates may be particularly beneficial in this respect. Concretes containing some limestone aggregates can have thermal expansion coefficients which are only two thirds of the value exhibited by concrete containing siliceous gravel aggregates.

Although not often used in large volume pours, lightweight aggregate concrete also has a relatively low thermal expansion coefficient.

The tensile strain capacity ε_{tsc} of concrete (and hence its crack resistance) is also significantly affected by the choice of aggregate[3], with those aggregates exhibiting low values of α, also achieving concretes with the highest values of ε_{tsc}. Limestone is again the best option among the commonly used aggregates and siliceous gravel is the worst. Lightweight aggregate also offers significant benefits in relation to ε_{tsc}.

The combination of increased tensile strain capacity and low thermal expansion coefficient associated with the use of limestone (and lightweight) aggregate in concrete makes this aggregate type particularly suitable for large volume pours where cracking is to be avoided. Table 2 gives estimated limits on temperature differentials for concretes with defined values of α_c and ε_{tsc}.

Under conditions of sustained loading, the tensile strain capacity is marginally reduced by the use of ggbs or pfa in concrete. This is believed to be due to the lower creep associated with these mix types[26,35]. In the absence of specific quantitative data, it is impossible to give revised values of tensile strain capacity. However, based on limited observations of large volume pours, minimum levels of ggbs and pfa have been proposed[34] in relation to the pour thickness to ensure that the reduction in temperature rise more than offsets the reduction in strain capacity (see Table 5).

Table 5 *Recommended minimum levels of ggbs and pfa*

Pour thickness	Minimum percentage level of:	
(m)	ggbs	pfa
Up to 1.0	40	20
1.0 to 1.5	50	25
1.5 to 2.0	60	30
2.0 to 2.5	70	35

These figures are for guidance only, and it is recommended that further research be carried out in this area.

4.5.3 Summary

A successful concrete mix design for large volume pours should have the following characteristics:
- Appropriate workability and strength
- Cohesiveness and low bleed
- Extended stiffening time to minimise cold joints
- Low expansion aggregate such as limestone
- Minimum cement content compatible with strength and durability and the requirements for placing
- Low heat cement, probably based on blends of Portland cement with either pfa or ggbs.

5 Construction in deep lifts

This section provides guidance on the practical aspects of construction in deep lifts.

Whilst the normal construction requirements for producing a high quality structure apply equally well to deep lifts, there are also a number of aspects particular to this form of construction that must be considered.

5.1 CONSTRUCTION SEQUENCE

When deep lift construction is proposed, the sequence and timing of the individual pours must be considered both in terms of speed and economy and in order to minimise the risk of thermal contraction cracking. The sequence of pouring is thus influenced by the thickness of the section and the concrete mix design.

In thin sections containing concrete with a low heat generation (i.e. based on composite cements), the effects of restrained thermal contraction are small and construction based on pouring alternate sections (often called 'hit and miss') may be appropriate. This technique also has advantages in that the infill panel is supported by the two mature adjacent panels, thus reducing the need to prop the formwork for extended periods (see Section 5.2).

In cases where either the section thickness is high (above 500 mm) or the heat generation of the concrete is unavoidably high, the effects of restraint to thermal movement must be considered. In such cases, less end restraint to thermal contraction is offered by a sequential ('end-on') form of construction[7]. Base restraint is almost unavoidable if walls are cast onto a mature (cold) base slab and is best reduced by casting shorter lengths. The effects of base restraint can also be reduced by selecting low heat generating concretes, or concretes with a low coefficient of thermal expansion.

Restrained thermal contraction in thin walls (which are often constructed using the deep lift method) can lead to vertical cracking. Guidance on how best to avoid and/or control this is given in CIRIA Report 91[7].

5.2 FORMWORK

Formwork for deep lift construction must be carefully designed and the following factors must be considered:

5.2.1 Formwork

Construction in deep lifts when combined with highly workable concrete and efficient compaction will generate a considerable hydrostatic head and the formwork must thus be designed to resist the consequently high formwork pressures. Extended stiffening time resulting from the use of admixtures or composite cements (in order to produce a cohesive concrete, or prevent cold joints) will also contribute to increased formwork pressures.

Guidance on this subject is given in CIRIA Report 108[28] which, in addition to hydrostatic pressures, also takes account of the influence of the impact of fresh concrete dropped from the top of the form.

It is probable that formwork for deep lifts will be constructed from large panels (often prefabricated), as one of the prime reasons for choosing this construction technique is to avoid joints and achieve the resulting improvements in surface appearance. Particular attention must be given, therefore, to the joints between the panels. Grout seals between panels (particularly in the lower parts of the lift) must be efficient and capable of withstanding the expected concrete pressures. Failure of seals will result in extensive grout loss and possible honeycombing.

Formwork for deep lift construction is particularly sensitive to wind loading. During formwork erection care must be taken to ensure that the propping will resist overturning[45]. This is particularly important when a single form is being erected. Stability of the filled form is also critical.

In order to achieve alignment with subsequent adjacent lifts, propping should remain in place for extended periods until the concrete has achieved a strength greater than that normally required for formwork removal (see Section 5.6). This will ensure that wind forces do not cause deflections of individual lifts resulting in an uneven surface of the overall structure.

5.2.2 Access

Access to deep lifts is frequently restricted to the top opening of the form, which may often be very narrow. All placing and compacting activities must, therefore, be concentrated in this restricted area. Provision must be made for adequate working platforms to allow these activities to be carried out efficiently and safety.

Another aspect of formwork design is the possible need for access doors. This is considered further in Section 5.4. Such access doors must be designed with suitable dimensions to allow for concrete placing by pump, flexible hose or other methods. A large number of such access doors may be required to enable successful placing. During the construction of 15.5 m high columns for a power station in India for instance, access doors 400 mm × 400 mm were spaced at 2.25 m centres on opposite faces of the column[46] (see also Figure 11).

Provision of (temporary) access platforms may also be required at the location of the access doors and the question of sealing must also be addressed. As with the joints between formwork panels, the seals around access doors after they are replaced in the formwork must resist the pressure of overlying concrete.

Although it is probably not possible to prevent the outline of the access doors being visible in the concrete surface after removing the formwork, every effort must be made to ensure that the access doors are flush fitting with the remainder of the formwork.

If possible, access doors should be positioned in areas of the lift that will not be visible in the completed structure.

Consideration should also be given to the provision of access for assessing the concrete strength at the time of formwork striking. A number of non-destructive or semi-destructive methods are available[47, 48] now covered by British Standards[49]. Either direct access to the concrete surface or provision of a pull out insert is often required. Liaison between the formwork designer and the specifier will be required in order to plan the most appropriate number and location of these access points. Once again these should be located if possible in areas that will not be visible.

5.2.3 Effect of formwork on surface finish

Surface finish is an important feature of much deep lift construction and this can be significantly affected by the formwork.

- The formwork panels must be flat and flush with each other with no steps or gaps. Welding steel panels may introduce distortions that will be apparent when the formwork is removed.
- The formwork material and its interaction with the mould release agent must minimise blowholes. Although blowholes are usually concentrated at the top of the lift, they can occur throughout the height of the pour if the formwork materials are unsuitable or poorly covered in release agent. This is particularly noticeable with low workability or 'sticky' concrete mixes.

Figure 11 *Access doors in a deep lift wall panel (Courtesy Costain-Taylor Woodrow JV)*

Impermeable materials such as steel, plastics or resin-coated plywood are particularly prone to excessive blowholes as water and air moving away from internal vibration becomes trapped at the concrete-formwork interface. In contrast more permeable materials absorb some of this water and reduce blowholes. In a recent deep lift construction[50] where opposite formwork faces were constructed in steel and traditional plywood, the ply face produced a much better finish in terms of reduced blowholes.

Variations in the absorption capacity of formwork materials may cause colour variations in the concrete, the lower water content in the concrete surface giving a darker colour. The higher hydrostatic pressure of fresh concrete on the formwork near the base of the lift may also increase formwork absorption and hence darken the colour.

Controlled Permeability Formwork (CPF) has been shown to be an effective means of eliminating blowholes in vertically formed surfaces[51]. Although the surface is darkened as a result of a low water/cement ratio, the darkening is uniform due to the *controlled* permeability of the surface. This technique has been successfully applied to deep lift construction (Figure 12).

5.3 REINFORCEMENT

The primary purposes of reinforcement are to ensure structural performance and to control cracking. This also applies to deep lift construction.

The crack control function is particularly important in deep lift construction where both joints and cracks are to be avoided. Reinforcement design should consider the possibility of thermal contraction cracking and design in accordance with BS 8007, Design of concrete structures for retaining aqueous liquids[9], may be appropriate.

Figure 12 *Controlled permeability formwork in deep lift construction (Courtesy Dupont-Zemdrain)*

There are also certain considerations specifically associated with deep lift construction.

- Congested reinforcement – the restricted access for placing and compacting concrete in deep lifts requires particular attention to the difficulties of properly compacting the concrete around the reinforcement. Heavily congested reinforcement may prevent proper compaction and may preclude deep lift construction, even with high workability concretes containing small aggregates. If reinforcement congestion is unavoidable, the possibility of using formwork access doors or an alternative construction technique should be considered.
- Reinforcement positioning – to enable concrete to be placed in deep forms without segregation, placing via pump lines, flexible trunking or tremie tubes are often required. To facilitate this, it is important that the reinforcement is positioned in such a way as to allow access for the placing tube.
- Consequently, those sections where most of the reinforcement is located close to the formwork faces are best suited to deep lift construction. Alternatively, in wide deep lifts, where the internal reinforcement is more accessible, reinforcement across the centre of the section can be fixed just ahead of the level of concrete.
- Even if placing from the top of the form via a skip or chute is unavoidable, minimising the number of cross links between the formwork faces can be an important step in reducing the risk of separation of aggregate and mortar during the drop to the base of the form.
- As with many aspects of specialised construction, liaison between designer and contractor regarding the preferred construction technique can reduce the number of problems on site, if reinforcement is specifically designed for deep lift construction.
- Maintenance of cover – spacers used to maintain concrete cover in deep lifts must be properly fixed as no adjustments can usually be made during concreting. Spacer fixing should be resistant to pressure from the overlying concrete and impact from concrete falling from above.
- Suitable access for fixing the reinforcement.

5.4 CONCRETE PLACING

One characteristic of deep lift construction, is restricted access for placing concrete and it is often impossible to avoid placing concrete from the top of a deep narrow form. Free fall of concrete should always be avoided as it often leads to segregation of the concrete and possible honeycombing, poor compaction and poor surface finish. Consequently a number of special placing techniques have been adopted for deep lift construction (see Figure 13).

5.4.1 Pumping

Properly proportioned concrete mixes can be pumped over considerable distances (both horizontally and vertically). When combined with an articulated placing boom, this can also reduce the requirements for access platforms and scaffolding.

The concrete pump hose can in certain circumstances be used to place concrete directly in deep lifts. However, the length of the hose is sometimes restricted to about 4 m for 125 mm dia hose and 3 m for 150 mm dia hose[20]. The diameter of the hose will be controlled by the maximum aggregate size in the concrete.

If, however, the section dimensions and reinforcement design allow the pump hose to be inserted into the form, it should be lowered as close as possible to the base of the lift and raised slowly as concrete placing proceeds. The hose should also be positioned vertically down the centre of the form to avoid directing concrete against formwork or reinforcement that could lead to segregation. A number of separate insertions may be required in order to distribute the concrete uniformly.

Pumps can also be used in conjunction with the following placing methods as a means of transporting concrete to the top of the formwork and have the advantage of reducing access requirements for placing.

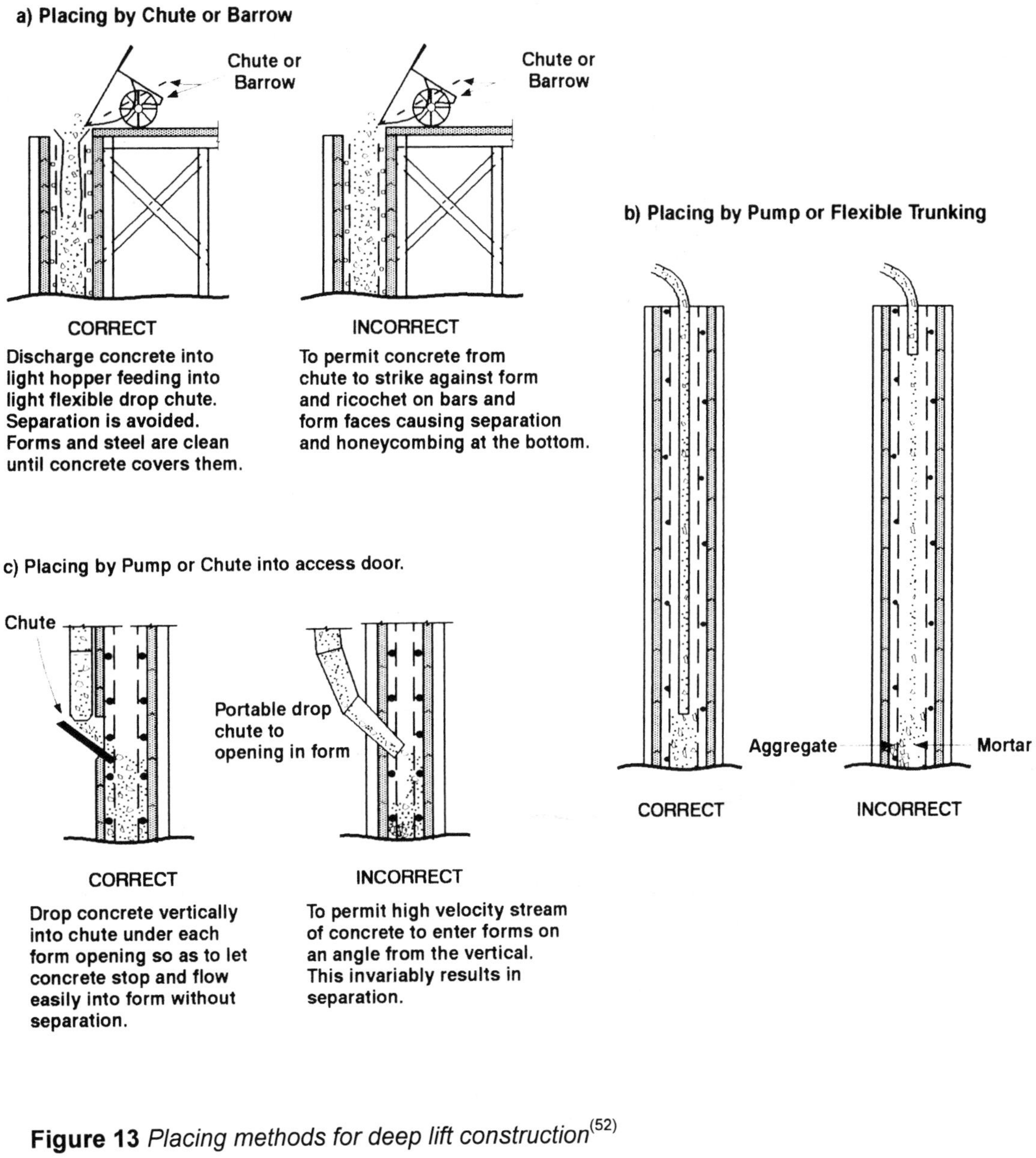

a) Placing by Chute or Barrow
Chute or Barrow
Chute or Barrow
CORRECT
Discharge concrete into light hopper feeding into light flexible drop chute. Separation is avoided. Forms and steel are clean until concrete covers them.
INCORRECT
To permit concrete from chute to strike against form and ricochet on bars and form faces causing separation and honeycombing at the bottom.
b) Placing by Pump or Flexible Trunking
Aggregate
Mortar
CORRECT
INCORRECT
c) Placing by Pump or Chute into access door.
Chute
Portable drop chute to opening in form
CORRECT
Drop concrete vertically into chute under each form opening so as to let concrete stop and flow easily into form without separation.
INCORRECT
To permit high velocity stream of concrete to enter forms on an angle from the vertical. This invariably results in separation.

Figure 13 *Placing methods for deep lift construction*[52]

5.4.2 Trunking and tremie placing

Even if the pump hose cannot be inserted directly into the form, lightweight flexible hose (often called 'elephant trunks') can be used, with concrete being fed into a hopper at the top of the hose. Trunking is often made of a flexible material that collapses (flattens) when suspended. This flattening slows the concrete free fall and reduces segregation.

The trunking should be positioned in the centre of the form (reinforcement permitting) and lowered to the base of the lift at the start of the placing and raised as the level of concrete rises. Trunking diameter should be at least six times the maximum aggregate size.

In large deep lift pours, a number of lengths of trunking may be required in order to distribute the concrete uniformly. It is important to minimise the distance through which concrete must be moved purely by vibration as this may also lead to segregation.

While flexible hose or trunking is usually preferred, rigid (tremie) tubing has also been used[46]. The considerations described above will still apply to rigid piping although heavy lifting plant may be required. As the tremie pipe is raised, individual sections can be removed to maintain the level of the hopper and reduce the load on the lifting equipment.

If it is physically impossible to insert a pipe or trunk to the full depth of a section, the risk of segregation (as a result of free fall of concrete from the top of the form) can be reduced by using a hopper and flexible drop chute that initially directs the concrete down the centre of the form avoiding the reinforcement and formwork faces. Concrete can be placed in the hopper using a pump, skip or barrow. It has been suggested[52] that the slump of concrete should be reduced gradually as the level of concrete in the form rises. This was in order to counterbalance the effects of water gain in the upper parts of the lift. However, this practice is not recommended as the degree to which slump must be reduced is a matter of arbitrary judgement, further complicated by the influence of admixtures rather than water content to control slump.

5.4.3 Placing via formwork access doors

Another technique that has been used to reduce the depth through which concrete falls during deep lift construction, is to place the concrete through doors in the formwork at various heights (see Section 5.2.2).

Concrete can be introduced into the form by this method using a pump line or via a chute or hopper (see Figure 13). Which ever method is used, it is important to ensure that the stream of concrete is not aimed directly at the opposite face of the formwork as this will result in segregation.

As the level of concrete rises to the level of the access door, the door is quickly replaced and sealed and concrete placing is transferred to an adjacent (horizontal or vertical) door. The uniform distribution of concrete is dependent on the spacing of the formwork access doors.

5.5 COMPACTION

Two particular problems affect compaction of concrete in deep lifts: firstly, the difficulty of inserting vibrators into deep narrow forms and controlling their position; and secondly, the problems of observing the concrete as it is vibrated, in order to gauge when compaction is satisfactory.

The use of internal (poker) vibrators with long air lines is common for compacting concrete in deep lifts. However, if the vibrator is controlled from the top of the form and just dropped into the concrete at the base of the form, it is difficult to control its position and areas of poor compaction may result. More than one poker may often be required in order to produce efficient compaction. Increased numbers of vibrators are essential if the design of the reinforcement does not permit the poker to be raised easily and repositioned. The workability and consistance of the concrete will also influence the number and spacing of the vibrators. The maximum diameter of the poker will, of course, be controlled by the section dimensions and reinforcement spacing, and pokers of different diameters may be needed for different sections of the lift. In some circumstances, it may be possible to prefabricate a jig for holding the pokers in such a way as to allow more accurate positioning and controlled upward movement. Vibrators should be inserted vertically and allowed to sink under self-weight. The radius of action (or area of influence) of the vibrator should preferably be larger than the wall thickness in narrow sections[53] (see Table 6).

Table 6 *Typical radius of action of poker vibrators*[53]

Vibrator dia (mm)	Radius of action (mm)
20-40	80-150
30-60	130-250
50-90	180-360
80-150	300-510
130-180	400-610

Vibrators should be inserted at a spacing generally around 1.5 times the radius of action so that the areas affected by vibration overlap. Vibrators should always be withdrawn slowly from the concrete.

If possible, contact between the vibrator and the formwork or reinforcement should be avoided.

External form vibrators are generally less efficient than immersed poker vibrators and also require modifications to the formwork to allow for fixing and the additional imposed loads. Excessive external vibration is also thought to cause an increase in the number of blowholes and sand runs in the formed face.

Although revibration is often suggested as a means of eliminating the effects of plastic settlement (a potential problem for deep lift construction), it should be noted that, as access is usually restricted to the top of the formwork, it is unlikely that revibration will eliminate settlement cracking at any significant depth within the lift. However, settlement cracking on the top surface may be prevented. This highlights the need to use a cohesive low bleed concrete mix for deep lift construction.

Despite these potential disadvantages, external vibrators have been used for deep lift construction particularly in areas of congested reinforcement where the use of internal vibration is difficult.

During deep lift construction of a reactor cell in a chemical works[54](see Figure 14), internal poker vibrators were supplemented by air powered external vibrators. As the level of concrete rose, the external vibrators were also moved upwards (in 1 m increments) and operated in short bursts.

Internal vibrators should also be moved upwards in line with the rising concrete level.

The main problem in assessing the effectiveness of the compaction process, is lack of visibility, particularly at the base of the lift. In some cases[53], powerful halogen lights have been lowered into the formwork. Barrel inspection lights have also proved effective. The use of video equipment may help in some circumstances. Formwork access doors are also useful, both for inserting poker vibrators into the formwork at lower depths in the lift and for observing the movement of the fresh concrete during vibration. Positioning the access doors close to areas of congested reinforcement will also enable smaller diameter pokers to be inserted into the concrete in these areas to improve compaction around the reinforcing bars.

Proper access for poker operators must be provided as this is a much more difficult operation in deep lift construction than in other forms of construction. Concrete pumps and articulated placing booms can be used as a means of reducing the number of placing personnel required on the access platforms, thus allowing the poker operators adequate room for manoeuvre.

5.6 FORMWORK STRIKING AND CURING

There are a number of potential problems when formwork is removed from deep concrete lifts, the most serious being collapse of the structure or excessive deflections. Factors to be considered are:

- collapse
- excessive deflection
- mechanical damage
- freeze/thaw attack
- thermal cracking

The minimum striking time is usually based on an assessment of the in-situ strength of the concrete.

A number of techniques for assessing the concrete strength prior to striking the formwork are discussed in CIRIA Report 73[47, 48]. Although a minimum in-situ strength of 2 MPa is often specified for resistance to mechanical damage or frost attack[45], a higher strength (in some cases up to 20 MPa[54]) is recommended for deep lift construction. This is primarily to ensure that the lift is able to withstand wind loads without causing deflections that will produce misalignment with adjacent panels.

In massive deep lift construction, striking times may be controlled more by thermal requirements than concrete strength[7].

Following the removal of the formwork, it is important that the concrete surface is protected from moisture loss (particularly in thin sections) in order to promote continued cement hydration leading to strong durable concrete.

This may be achieved in a number of ways, including:

- Covering the surfaces with wet hessian and then with polythene
- Application of an efficient membrane forming curing compound.

Recommended minimum curing periods are given in BS 8110[24]. The choice of curing method will be influenced by practical considerations and any subsequent treatment to be applied to the concrete [55].

a) During construction

b) Completed cells

Figure 14   *Deep lift construction of chemical works reactor cells*
(Courtesy Alfred McAlpine Construction Limited)

5.7 MONITORING AND ACTION TO BE TAKEN IN THE EVENT OF NON-CONFORMANCE

The action to be taken in the event of non-conformance with specific strength requirements is summarised in Section 2.7. Whilst monitoring of deep lifts per se is not an essential requirement of successful construction, when the section is of large volume, temperature and strain measurements may be required to demonstrate conformance with specified limits. This subject is covered in greater detail in Section 6.

However, in terms of non-conformance, two particular aspects must be considered:

* visual appearance
* cracking.

Following the removal of formwork, a general inspection of the appearance of the concrete is required. This should look for the following features:

* area of poor compaction or honeycombing
* areas of grout loss
* excessive blowholes
* lips or misalignment between formwork panels
* sand runs and settlement cracks
* cracking
* cover.

With the possible exception of cracking, poor surface appearance is the responsibility of the contractor. Remedial action, i.e. bagging in, grinding down lips or patch repairs of honeycombed areas, should be evaluated in the light of the consequences of the defect, i.e. structural or aesthetic, and treated accordingly.

The subject of cracking however, is more contentious. Many structures constructed in deep lifts are also designed as water-retaining structures. In such cases, the reinforcement design is based on limiting crack widths to an acceptable mean value[9]. If cracking is observed in the section, a number of questions must be addressed:

* Does the observed extent of cracking exceed the design value?
* What assumptions has the designer made about the concrete performance in order to design the reinforcement to control cracking?
* Were these assumptions passed on to the contractor, i.e. in terms of temperature limitations and specifications of aggregate type or limits on α_c and ε_{tsc}?
* Did the contractor comply with the specified requirements?
* What are the consequences of the cracking?
* What remedial measures should be taken?

The first step should be to measure the crack widths using a crack width microscope or comparator. It should be noted, however, that difficulties often arise both in obtaining a representative value and in agreeing acceptable limits. In particular, the following questions need to be addressed:

* Is the design crack width the value at the steel, or the value at the surface?
* If the crack tapers, where should the measurement be taken?
* How many measurements should be taken to provide a representative value?

In the absence of definitive guidance it is proposed that a mean crack width is derived, taking into account all cracks within a cast section. For each individual crack, the value at the widest point should be taken. The calculated mean value should then be compared with the maximum design value to establish conformance.

If the crack width is 0.2 mm or less, there is a high probability of self-healing, particularly in a moist environment[56]. If the mean crack width exceeds the maximum design value, remedial work (i.e. resin injection) should be considered.

Assuming that the design method is correct, then excessive cracking will only occur if the magnitude of restrained thermal contraction has exceeded the estimated value. This could arise if:

- The actual value of ΔT exceeded the design value. If a limit was specified, then the contractor would have to accept responsibility.
- The actual value of thermal expansion coefficient, α_c, of the concrete was higher than the estimated value. It is rare for the designer to include an upper limit for α_c in the specification although, occasionally, the aggregate type is specified, e.g. limestone. Even in the latter case, however, it must be recognised that α_c can vary. For example siliceous limestone aggregate may yield a higher α_c than achieved by the use of calcareous limestones. If the designer has simply underestimated α_c, then the contractor cannot be held responsible if he has complied with all other aspects of the specification.
- The restraint is higher than estimated. Both BS 8007[9] and CIRIA Report 91[7] assume effective restraint values of 0.5 which include factors to take account of creep. In reality, the restraint will vary[3] depending on the size and stiffness of the new concrete in relation to its immediate surroundings, i.e. adjacent concrete, rock etc, and the extent to which heat transfer aids compatibility of thermal strains. True restraint is difficult to estimate reliably, although the ACI[10] offers a more detailed evaluation than any of the UK methods. It is likely that in many cases, excessive cracking is due to an underestimate of restraint, and this will continue to be the case until more data are obtained from structures.

In such cases, disputes can only be resolved by obtaining in-situ strain and temperature data to measure restraint directly, but this cannot be done after the event. Hence on large contracts, strain measurements should be considered in the first few pours to define the safety margins, and the scope for changing bay lengths as the contract proceeds.

Thus, while the responsibility for remediation would lie with the contractor in the event of a clear failure to comply with the specification, non-conformance must be considered in relation to assumptions made by the designer, and the extent to which the contractor was made aware of those assumptions. In cases where conformance is achieved, but the design assumptions are not valid, then the designer must take full responsibility for remediation.

This highlights the need for closer cooperation between designer and contractor in order to avoid potential problems. In particular, any design assumptions such as expected temperature rise or thermal expansion coefficient, must either be reflected in specific requirements in the specification or be discussed with the contractor prior to construction.

6 Large volume pours

This section provides guidance on the practical aspects of construction of large volume pours and means for the control of early age thermal cracking.

6.1 REINFORCEMENT DESIGN AND DETAILING

When designing reinforcement for large elements which have the potential to be cast as single pours, in addition to the normal structural requirements, the designer must also consider the following:

1. Control of early age thermal cracking (Section 2)

2. Avoidance of steel congestion
   - to enable access for internal (poker) vibrators through the top mat steel. Pokers vary in size from 25 mm to 75 mm in diameter, with the larger sizes being most appropriate in large volume pours. The pokers should be lowered vertically at about 0.6 m centres.
   - to enable proper compaction around the steel and at faces and corners.

6.2 FORMWORK

As large volume pours are generally relatively low height (a few metres at the most), there are no special requirements for the formwork, other than to provide vents through which rainwater can be expelled as it is driven ahead of the advancing concrete face. The vents must be located after defining the sequence of concreting. When the finish is important, guidance is given in Section 5.2.3.

When adjacent pours are to be cast, it is common to use expanded metal formwork at joints. This is supported during casting, but the supports can be removed within a few hours due to the accelerated strength development in the large pour.

6.3 CONSTRUCTION PROCEDURE TO MINIMISE RESTRAINTS

When a large element is being cast as a number of individual large volume pours, the sequence of construction should aim to minimise restraints to reduce the extent of cracking. This involves the avoidance of infill pours, as far as is practically possible. End-on-end construction provides a free end to accommodate movement, thus minimising the effect of restraint to bulk thermal contraction.

With regard to the delay between adjacent casts or lifts, the following general rules apply:

1. The time between adjacent strips or lifts should be minimised, as the principal restraint acts along the direction of the joint (Figure 15a). If the previous pour is still warm, temperature (and hence strain) differentials between new and old concrete will be minimised as the two pours contract together.

2. The time between end-on elements should be maximised, as the principal restraint acts perpendicular to the joint (Figure 15b). This eliminates contractions in the old pour prior to casting the new element.

3. To avoid excessive restraint in either direction the preferred shape for a given volume of concrete is as close to square as possible. This avoids the need for specific timing between adjacent elements.

4. If a source of local high restraint falls within the area of the proposed pour, it may be prudent to introduce a construction joint close to it. The gap, to be filled once the large pour has fully contracted, should be as small as practically possible, taking account of the access needed to introduce and remove the stop end, and to prepare the joint surface before the infill is cast. While the temperature cycle may be estimated, it is better to monitor temperatures in the initial pours or, preferably, in trial pours and to programme the casting on this basis.

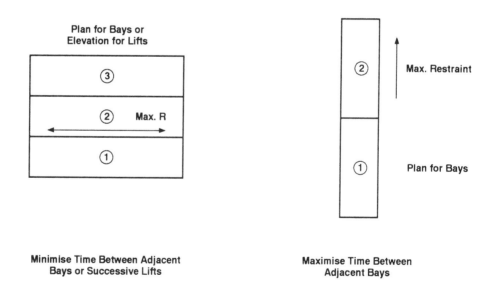

Figure 15 *Pour configuration as it affects restraint and delay between casting adjacent bays*

In relation to restraint, it is also important to recognise the following:

- In very thick sections, external restraint is likely to be much less significant, as large forces can be generated even though the concrete may be relatively young. In such cases, cracking is more likely to result from internal restraints caused by temperature differentials.
- In sections which are relatively thin, and cast against existing mature concrete, external restraint will be predominant. The influence of internal restraints are diminished as the external restraint prevents differential internal strains developing.

In practice both internal and external restraint exist simultaneously, although in many cases one or other is predominant. Where combined restraints exist, the effects are superimposed[11], as shown in Figure 16 for a thick wall cast onto a rigid foundation. This demonstrates the difficulty in predicting restraints, and the level of simplification which has necessarily been adopted by BS 8007 and CIRIA Report 91. The latter acknowledges that during heating, external restraint will cause a reduction in the tensile stresses in the surface zone as bulk expansion is resisted. It recommends that, where external restraint exists, precautions need not be taken to minimise differential temperature as this is likely to increase the extent of cracking.

Because of the inherent complexity of the problem, each case must be considered separately. However, the following broad guidelines may be followed:

1. For elements of large cross section, where external restraint is low, steps should be taken to limit temperature differentials as defined in Table 2 for different mix types.

2. For thin elements (<500 mm) subject to high external restraint, temperature limits in excess of those given in Table 2 may be tolerated, subject to an analysis of the likelihood and/or extent of cracking.

Reasonable estimates of external restraint can be derived using the ACI method described in Section 2.

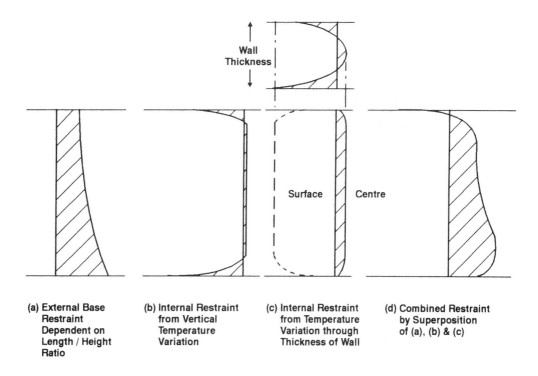

Figure 16 *The superposition of internal and external restraint*[11]

6.4 PLACING METHODS

The four principal methods of concrete delivery into large volume pours are:

- skip
- pump
- chute
- conveyor

The method selected will depend on the required rate of placing, the ease of access and the plan area of the pour. When the plan area is relatively small, chutes provide an effective method of placing, but will only be successful for large volume placements in which high workability concrete is used. Using this system with superplasticised flowing concrete, placing rates of 80 m^3 per hour have been achieved, 400 m^3 in five hours (Figure 17).

Conveyors provide access over greater distances, but are most successful for low workability concrete. High delivery rates can be achieved (60 m^3/hr), but this is dependent on the mix type, being higher for low slump concretes.

However, for large area, large volume pours, mobile boom pumps provide the greatest degree of flexibility combined with very high placing rates, typically of the order of 40-50 m^3/hr, but with reported throughputs approaching 100 m^3/hr for individual pumps. This compares with typical rates of 8-12 m^3/hr achieved using crane and skip.

Figure 17 *Placing a large volume pour using high flow concrete and chutes (Courtesy Taylor Woodrow Construction)*

The volume of concrete, the depth of the pour and its plan area all influence the sequence of placing and the number of placer units. Large volume pours are best placed in successive layers. Each layer is normally up to about 600 mm thick. Shallower layers reduce the risk of poor compaction, but layers up to 600 mm thick can be properly compacted with sufficiently large vibrators.

The principal objective is to achieve a monolithic element with no planes of weakness, hence the construction sequence must be designed to minimise the period of exposure of the newly placed surfaces. This will lessen the effects of warming in hot weather, and reduce the effects of surface damage during rainfall.

For long, narrow pours, the step method is most effective in this respect and involves commencing successive layers after previous layers have been compacted sufficiently far ahead. Fitzgibbon[18] proposed that the distance between the progressing faces could be ten times the layer thickness, i.e., up to about 5 m. The ACI[4], however, recommend that this distance should not exceed about 1.5 m although this applies specifically to low slump concrete. The actual distance used will depend on the thickness of each layer and the stiffness of the fresh concrete. For thin layers (300 mm) of low slump concrete, 1.5 m is more appropriate. When thicker layers (500 mm) of higher slump concrete are used, 5 m may be necessary to avoid 'sagging'of the front.

To advance the 'stepped' front at a reasonable rate several placers are necessary, but if the pour is deep enough to complete in, say, three layers, the length of the section becomes unimportant.

When several placers are used, the options are numerous, and often involve dividing the pour into discrete segments, each of which is cast in layers.

For circular elements, there is often the temptation to start at the centre and work towards the circumference. However, this makes it increasingly difficult to maintain a 'live' front, and working from side to side may offer better control.

6.5 COMPACTION

For large volume pours, the only practical means of compaction is by the use of immersion (poker) vibrators. Recognised good practice involves placing the poker vertically into the concrete and at about 0.6m centres. At each location, the time required to ensure expulsion of air will depend on the workability of the concrete.

A simple nomogram derived from data in Orchard[57] is shown in Figure 18, which enables typical rates of compaction to be estimated. For a 75 mm slump concrete, an output of 20 m^3/hr per poker can be expected. This is consistent with ACI recommendations[53] which suggest rates of 6-20 m^3/hr for concretes with a slump less than 75 mm.

Where plastic settlement cracks are observed, it is acceptable to revibrate the surface layer. Concrete Society Report No. 22[31] confirms that this need not be damaging, provided that it is applied neither too soon, allowing further bleed, nor too late, causing disruption of partially 'set' concrete. Site trials are recommended to define the window within which to work.

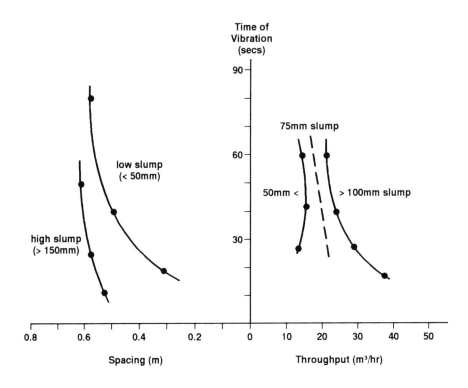

Figure 18 *Nomogram relating concrete workability, poker spacing and throughput*

6.6 FINISHING

Large volume pours will generally be either trowelled or floated. Normal practice will apply, but it must be recognised that the concrete mix may have been designed to have an extended stiffening time to avoid cold joints. For this reason, there may be a longer than usual delay before initial stiffening has taken place and the moisture film has disappeared from the surface.

6.7 THERMAL CONTROL

The control of temperature rise and temperature gradients is often a key feature of large volume concreting and various precautionary measures can be taken to minimise the risk of early age thermal cracking, or to limit its extent as follows:

- Select a low heat generating mix by minimising the cement content, within the constraints of the specification and requirements of the fresh concrete
- Use composite cements (i.e. blends of PC with pfa or ggbs) to reduce the rate of heat evolution and the peak temperature rise
- Select aggregates with a low coefficient of thermal expansion
- Select aggregates which result in concrete with a high strain capacity
- Reduce the concrete mix temperature by cooling of the component materials or the use of ice with the mix water
- Use liquid nitrogen to cool the fresh concrete
- Cool the concrete in-situ by the use of embedded cooling pipes
- Use insulation to increase surface temperatures and to minimise thermal gradients.

6.7.1 Low heat concretes

Low heat means a low cement content mix or the use of a cement with low heat generating characteristics.

There is now considerable evidence supporting the beneficial use of composite cements (Portland cement blended with either pfa or ggbs) in large volume pours. Results given in Section 4.5.1 show the extent to which reductions in temperature rise can achieved. However, as cements may vary between sources, tests are recommended to determine the temperature rise for the selected mix. In this respect Coole[38] has reported that isothermal calorimetry cannot predict the effect of all cement types, being deficient for blends with ggbs in particular. Adiabatic, or near adiabatic (heavily insulated block) tests provide data which are of more direct relevance. On site, a typical test block would involve a 1 metre cube insulated with 50 mm of expanded polystyrene, but may be of any dimension which simulates pours to be cast.

Reductions in cement content may be less easy to accommodate, although in Section 4.5.1 it is proposed that the benefit of the enhanced in-situ strength of pfa concretes be used to permit a reduction in γ_m, the materials safety factor.

The use of microsilica (also known as silica fume) also enables the cement content to be reduced while maintaining the strength[32]. While the microsilica itself makes a contribution to heat evolution, the effect on strength is greater with an overall benefit in terms of reduced temperature rise[58].

Increasing the maximum aggregate size will also enable a reduction in cement content and this is often proposed. However, there is also a change in strain capacity resulting from the use of larger aggregate/lower cement content mixes (Figure 7) and the detriment appears to be greater than the benefit of reduced heat.

6.7.2 Aggregate type

As discussed in Section 4.5.2, those aggregates which have a low coefficient of thermal expansion also exhibit higher strain capacity, and considerable benefits can be achieved in terms of relaxation of allowable temperature differentials by judicious aggregate selection. Although for most contracts the locally available aggregates can be accommodated in the design process, it may be appropriate for critical structures, e.g. where either radiation shielding or containment is a principal performance criterion, to use aggregates which will minimise the risk of cracking. As discussed in Section 4.5.2, of the naturally occurring aggregates limestone will produce concrete with the greatest resistance to cracking, while siliceous gravel is least resistant. Lightweight aggregate concrete outperforms all naturally occurring aggregates in terms of low α_c and high $\varepsilon_{tsc,}$ but also requires a higher cement content for the same grade, which partially offsets some of this benefit.

Acceptable limiting temperature changes and differentials are given in Table 2 which illustrate the significance of aggregate type.

6.7.3 Reducing the mix temperature

Reducing the mix temperature has a number of benefits:

- It reduces the rate of workability loss
- It delays the stiffening time
- It slows down the rate of heat generation
- It reduces the peak temperature, and thus the temperature drop from peak to ambient.

The most common method for cooling the mix involves cooling one or all of the individual mix constituents. As aggregates comprise the largest single component, cooling them will have the greatest effect on the concrete mix (except where ice is used). This can be achieved practically by:

- Shading aggregate stockpiles, to prevent solar gain
- Sprinkling the stockpiles with water (preferably chilled).

Other more extreme measures include immersion in tanks of chilled water, spraying chilled water on aggregate on a slow moving belt, or blowing chilled air through the stockpiles[59]. When the aggregates are cooled with water, this water must be taken into account during batching, by adjustment to the added mix water.

An alternative method is to cool the aggregate using liquid nitrogen (LN$_2$). This process has been used in Japan[60] to cool sand to -140°C and achieve a reduction in mix temperature of about 10°C.

While the aggregate constitutes the greatest mass in the mix, the water has the greatest heat capacity, and hence cooling efficiency. The specific heat of water is about five times that of the aggregate and cement (4.18 kJ/kgK compared with 0.75 kJ/kgK respectively). In addition, water is much easier to cool and the temperature can be controlled more accurately. As it is practical to cool water to about 2°C this is a very effective method of cooling the mix.

For very effective cooling, ice can be used. The latent heat of ice is 334 kJ/kg, and the heat absorbed by 1 kg of melting ice is equivalent to cooling 1 kg of water through about 80°C or 1 kg of aggregate through 445°C. Hence, a relatively small volume of ice can have a significant cooling effect. Ice is usually added to the mix in crushed or shaved form as it is important to avoid incorporating larger fragments that melt slowly leading to the formation of voids in the hardened concrete. To obtain an indication of the requirements for cooling the individual mix constituents, the following general rules may be applied[61]. To cool the concrete by 1°C requires that:

- the aggregate is cooled by 3°C
- the mixing water is cooled by 7°C
- 7 kgs of mixing water is replaced by ice.

A nomogram illustrating the effect of cooling the various mix constituents is given in Figure 19. This has been developed for a specific mix to provide a rapid means for identifying what steps are needed in specific cases. The nomogram is based on the simple method of mixture as follows:

$$\text{Concrete temperature} = \frac{0.75\,(T_c M_c + T_a M_a) + 4.18\,T_w M_w - 334\,M_i}{0.75\,(M_c + M_a) + 4.18\,(M_w + M_i)} \tag{6}$$

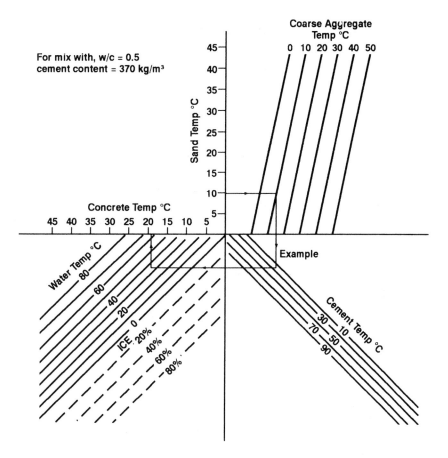

Figure 19 *Nomogram for estimating concrete mix temperatures*

where T is temperature, °C and M is mass, kg/m^3 and the subscripts c, a, w and i represent cement, aggregate, water and ice.

Where greater accuracy is required for a mix of different proportions, this equation can be used.

Recent developments with the use of liquid nitrogen (LN$_2$) now enable the mixed concrete to be cooled on site. The method involves spraying a mist of LN$_2$ (which has a boiling point of 77K (-196°C) into the mixer at a controlled rate. This is achieved with a customised lance which is inserted into a mixer truck.

The latent heat of boiling of the LN$_2$ is 199 kJ/kg, about 60% of that of melting ice. With 100% efficiency, about 12 kgs of LN$_2$ is, therefore, needed to cool 1 m^3 of concrete by 1°C. In practice, the efficiency is not much less than 100%. During construction of an X-Ray facility for Maidstone Hospital[62] about 1 tonne of LN$_2$ was used to cool a 6 m^3 mixer load through about 13°C. This is equivalent to about 13 kgs of LN$_2$ per °C change in temperature per m^3 of concrete (or 16 litres/°C/m^3). A similar rate of consumption of 15 litres/°C/m^3 was recorded during construction of the Faro Bridges in Denmark[63].

6.7.4 Embedded cooling pipes

When the specification prevents the use of concrete with low heat generating characteristics an alternative means for reducing temperature is the incorporation of an embedded cooling system. This has the advantage that the system can be designed to accommodate any mix type. The cooling system must be designed to remove heat at the required rate without inducing excessive internal temperature differentials. For this reason, plastic pipes may be preferred to metal pipes as the heat flow into the coolant is limited by the conductivity of the pipe itself.

A method for designing a cooling system is given in ACI Report 207 4R-80[59]. Typical pipe spacings are likely to be of the order of 1m in large volume pours with relatively low heat generating capacity. In elements cast using high grade structural concrete, closer spacing, of the order of 400-500 mm may be necessary.

The location of internal cooling pipes is unlikely to coincide with the reinforcement, as the latter is generally concentrated near the surface. However, there may need to be some collaboration between the designer and the contractor when this approach is adopted. This technique is often used to construct water-retaining structures. For example, in Germany[64], a rainwater reservoir was cast as a single element using this technique to control the thermal stresses. The 2 m thick base (plan area 27.4 × 28.2 m) and the 1.2 m thick, 12 m high walls were continuously cast (1100 m^3 of concrete) over a 32-hour period to avoid joints which were prohibited by the specification.

6.7.5 Insulation

Insulation is used to minimise temperature differentials by enabling the surface temperature to increase. It should be restated, however, that where the external restraint is predominant, the temperature differentials required to cause cracking are increased and insulation may be unnecessary. Furthermore, the use of insulation, by increasing surface temperatures in such cases will increase the risk of surface cracking. However, when appropriate, insulation can be achieved by various methods, including:

- maintaining the formwork in place if it has insulating properties e.g. plywood or steel backed with expanded polystyrene
- foam mats, blankets and quilts
- soft board
- sand on polythene
- tenting
- ponding

The method adopted will depend on whether the surface is vertical or horizontal, the period for which the insulation must maintained, and the access required to the surface.

Guidance on the minimum period for insulation in relation to the minimum pour dimension is given in Reference[3] and is reproduced in Table 7. For very thick elements, the insulation may need to remain in place for up to three weeks.

Table 7 *Minimum periods of insulation to avoid excessive temperature differentials*

Minimum pour dimension (m)	Minimum period of insulation (Days)
0.5	3
1.0	5
1.5	7
2.0	9
2.5	11
4.5	21

When using insulation care must be taken to avoid its early age removal, as this can generate thermal differentials which are much worse than might have occurred if no insulation had been used (Figure 20). Once the commitment to insulate has been made, it must be maintained until the temperature within the pour has cooled to a level which can accommodate the change which will occur when the insulation is removed. In this situation, the surface zone is heavily restrained by the bulk of the concrete and the expected temperature drop in the near surface zone must not exceed the limiting values given in Table 2 for a restraint factor of 1.

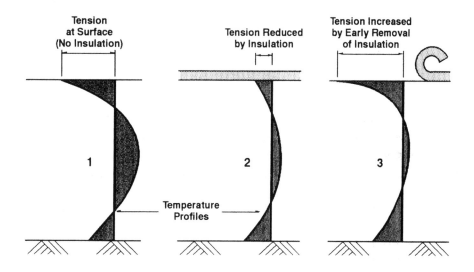

Figure 20 *Gradients developed in thick sections if insulation is removed too early*

6.8 CURING

The curing requirements for large volume pours are no different to those of concrete in general, with the exception of the thermal curing requirements to avoid excessive temperature differentials. However, the sealing of the surface associated with thermal curing which generally extends over several days, is likely to be more than adequate. In most cases, additional curing beyond the end of the thermal curing period will be unnecessary. Where further curing is required, guidance is given in Section 5.6.

6.9 MONITORING AND ACTIONS IN THE EVENT OF NON-COMPLIANCE

Temperature monitoring is generally required in order to demonstrate compliance with specified temperature limits. The simplest and most cost-effective method, particularly in very deep or thick pours, is to use embedded thermocouples. These are simple to install and to monitor. The output can be monitored with simple manual devices. On large projects, automatic logging systems can be set up.

In its simplest form, the monitoring will involve thermocouples at the centre of a section and close to the surface (about 10 mm). The maximum value at the centre and the difference between the centre and the surface will then be assessed against specified limits.

When the specification is more complex, for example, by limiting differences between a mean value and a total value, the designer must specify how the mean is to be determined. This involves defining how many locations through a section must be monitored, where the thermocouples are to be located, and how the results are to be analysed.

Where contracts specify the need to meet temperature limits, the contractor should obtain specific details from the designer on how compliance is to be demonstrated in relation to the location and number of monitoring points and the interpretation of the results obtained.

Full-scale mock-ups are very useful for providing data on temperature profiles and hence for identifying the precautions needed in terms of insulation, and increasing use is being made of computer models.

If the pour is a one-off, or if it is the first of many, pairs of thermocouples should be installed at critical locations to provide back-up in the event of a failure. The mix selection and adopted site procedures should be such that the temperature limits will be met. However, contingency plans should be defined which can be implemented when the results exceed 'action levels'. For example, if the allowable temperature differential is being approached, and is likely to be exceeded, additional insulation may be applied.

If the maximum allowable temperature is approaching its limit, no actions can be taken in thick sections unless embedded cooling pipes are being used. However, this value is much easier to predict based either on the mix constituents and the mix temperature or, if available, on laboratory test results for adiabatic temperature rise[38].

If allowable limits on temperature differential are exceeded the following action can be taken:
- Inspect the surface for cracks after the pour has cooled back to ambient
- If the cracks are excessive, initiate remedial measures
- Consider the consequences of internal cracking – it is possible that this may occur even if there is no evidence of surface cracking. In this case, it is difficult to establish the extent of cracking without extensive coring and even this is unlikely to indicate crack widths. The value of carrying out a detailed investigation to find cracks which do not propagate to the surface is, therefore, still doubtful, except in circumstances where such cracks are clearly detrimental to the performance and safety of the structure.

If cracks are to be repaired, the method used will depend on whether the cracks are 'dead' or 'alive', and on the serviceability requirements of the structure. In a massive raft for example, surface cracks are unlikely to increase after the element has cooled and sealing will be primarily for durability. In this case injection grouting is most common, using an epoxy resin when the cracks are less than about 1 mm width. Procedures are defined in Reference 65.

For cracks which are likely to continue moving, it is usual to make provision for further movement to continue after repair. This may involve a typical detail as shown in Figure 21[66].

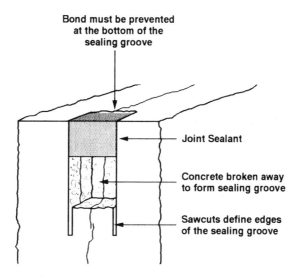

Figure 21 *Repair to moving cracks*[66]

The allowable maximum temperature is applied to limit the detrimental effect of temperature on the strength of concrete and sometimes to prevent delayed ettringite formation. If this value is exceeded, core testing will provide evidence of whether the deterioration in strength is excessive. BS 6089[67] provides guidance on the interpretation of in-situ strength, and it is advisable that the designer and the contractor agree the acceptance level of core strength before results are obtained.

7 Review

7.1 DEEP LIFT CONSTRUCTION

7.1.1 Advantages

The main advantage of construction in deep lifts is the reduction in the number of horizontal joints. This results in both an improvement in surface appearance and a reduced number of potential zones of weakness (or leakage in the case of liquid retaining structures).

From the construction practice point of view, deep lift construction also has certain advantages:

- Formwork and access scaffolding only needs to be assembled and erected once, thus reducing both installation and rental costs
- No joint preparation is required between small lifts
- Overall timescale of construction reduced.

Thus if the technique is both appropriate to the structure and successfully carried out, both structural and cost benefits can result.

7.1.2 Achieving successful deep lift construction

The main considerations for successful deep lift construction are proper planning, access and a concrete mix that is appropriate for use with this technique. The following points should be noted:

Specification

- The specification should be written with deep lift construction in mind
- The limitations on concrete mix proportions should be compatible with deep lift construction
- The contractor should be given maximum freedom on ways of achieving the specification requirements.

Construction

The concrete mix must be designed to be both workable and cohesive
- Techniques for placing and compacting concrete at the base of deep lifts must be decided in advance of construction and proper consideration given to access and construction sequence
- Formwork should be designed to resist high hydrostatic pressures and particular attention should be paid to sealing joints in the formwork
- Formwork should remain in place long enough to prevent excessive deflection leading to misalignment of panels
- A full-scale mock up of any particularly difficult section should be constructed and filled with concrete in order to identify and prevent any potential problems during construction.

7.1.3 Examples of deep lifts

Deep lift construction has been used successfully for a number of structures worldwide. Examples include:

- 14 m high retaining walls along the M20 motorway in Kent (including an intricately patterned surface finish)[68], where the wall was cast in a number of 5 m wide full height pours. Wall thicknesses were 1 m at the top widening to 2 m at the base.
- Fourteen 15.5 m high columns (2 m × 1.2 m in plan) were successfully cast in single lifts of 4-6 hours during the construction of the Trombay Power Station near Bombay in order to eliminate construction joints[46]. Concrete placing was achieved via a concrete pump and placer boom in conjunction with a tremie pipe. Formwork access doors were used to enable the concrete to be properly vibrated.
- A rain water reservoir in Berlin (base area 28 m × 27 m) was constructed with all four walls (1.2 m thick × 12 m high) being poured in a single lift of 1600 m^3 over a 36-hour period. Concrete was delivered to the form via pumps[64].

  Due to the width of the walls, operators with hand held poker vibrators could climb down into the formwork to compact the concrete.

  One interesting aspect of this structure was that cooling water pipes were also embedded in the walls to control the temperature build-up.
- The 12 m high 400-500 mm thick walls of two reactor cells at a UK chemical works were constructed in 25 pours over a 10-week period[54] using 'an alternate bay' construction sequence (see Figure 14). The pours were 6 m long and concrete was placed using a flexible hose rising at a rate of pour of 4 m/hr. The concrete mix was based on a composite cement (60%PC/40% ggbs) and a water/cement ratio of 0.45.

7.2 LARGE VOLUME POURS

7.2.1 Advantages

The principal advantages of large volume pours are as follows:

- Avoidance of construction joints, with their consequential implications on cost and timescale
- The construction of monolithic units.

7.2.2 Achieving successful large volume pours

The sizes of pours now being cast internationally, demonstrate that large volume pours are feasible and cost-effective. However, their successful completion relies upon:

- Recognition of the principal technical requirements at the design stage
- Cooperation between the designer, the contractor and the concrete supplier
- Adequate notice of start of casting given by contractor to all other parties
- Consideration of technical factors relating to the concrete including
  - stiffening time
  - plastic settlement
  - heat of hydration
  - selection of materials and mix design

- Planning to ensure that the concrete can be continuously delivered, transported, placed and compacted within the allocated timescale and in such a way as to avoid cold joints and other construction defects
- Contingency plans to take account of breakdown at any of the stages in the construction process, and the possibility of having to stop concreting, or to extend beyond the allocated timescale.

7.2.3 Examples of large volume pours

Numerous large pours have been cast worldwide and a limited review of some of the larger pours is given in Table 3. Some specific examples are given below:

- Thames Barrier cofferdams[17]
  A total of 6600 m^3 of concrete was cast over a three-day period. The mix used a 50:50 blend of PC and pfa to reduce heat of hydration, and a retarding admixture to achieve a delay in stiffening time of up to 36 hours. The 150 mm slump concrete was placed underwater using 300 mm diameter tremie pipes on a 7 m grid fed by mobile boom pumps.

- The 'Messeturm' mat foundation[6]
  The largest continuous concrete pour reported to date is the 5½ m thick 17 000 m^3 mat foundation for a 260 m high office tower in Frankfurt, Germany. The pour took 78 hours to complete using four mobile boom pumps; 90 readymix trucks from six plants were used for concrete delivery. Workers operated on a 12 hour shift basis, with 120 workers per shift.

- LINAC facilities[62]
  Although of relatively modest volume, liquid nitrogen was used to reduce the mix temperature to below 10°C for the construction of the joint free walls and roof of a radiotherapy centre at Maidstone, Kent. The 375 m^3 of concrete for the 1 m to 2.5 m thick walls was cast over a 13-hour period using a single mobile boom pump, using a low heat grade C30 mix comprising 70% ggbs, 30% PC and a 20 mm limestone aggregate. The formwork was removed after four to five days and no cracking was reported.

- Foundation mat, California, US[14]
  The most rapid reported rate of continuous output from a pump over a prolonged period was achieved during construction of a 5700 m^3 foundation mat in California. Five pumps achieved an average throughput of 496 m^3/hr (99 m^3/hr per pump) to complete the pour in only 11½ hours using a grade C20 mix with pfa.

8 Research Needs

Although deep lift and large volume pour construction has been successfully applied to a number of projects worldwide, certain aspects may benefit from further research. These include:

- Methods of evaluating and predicting the degree of cohesiveness, bleed and settlement of concrete mixes and setting appropriate limits for use in deep lift construction.
- Examination of the quantitative effects of vibration (both internal and external) on compaction, movement and segregation of concrete in deep lifts.
- In-situ evidence of the performance of different concrete mix types with regard to their risk of early age thermal cracking, including the effects on heat generation and strain capacity of composite cements.
- In-situ results of the effect of early age temperature rise on long-term strength development and durability and the influence of different mix types.
- Data to provide more reliable estimates of restraint factor for different pour sizes and configurations for use in the calculation of the risk of cracking and validation of existing predictive models.
- Measurement and interpretation of crack widths in relation to design and serviceability requirements.

References

1   BROOK, K M
    Placing concrete in deep lifts
    CIRIA Report 15, 1969.

2   BIRT, J C
    Large concrete pours – a survey of current practice
    CIRIA Report 49, 1974

3. BAMFORTH, P B
    Mass concrete
    Concrete Society, Digest No. 2, 1984

4. AMERICAN CONCRETE INSTITUTE
    Mass concrete
    Manual of Concrete Practice, 207, IR-87, 1988

5. ANON
    Consolidated Edison's record concrete floor pour
    Concrete Products, March 1969, 72 (No. 3) p46

6. ANON
    Pour sets world record
    ENR, Dec 1, 1988, p14

7. HARRISON, T
    Early-age thermal crack control in concrete – revised edition
    CIRIA Report 91, 1992

8. BAMFORTH, P B
    Early age thermal cracking in concrete
    Institute of Concrete Technology, Technical Note TN/2, 1982

9. BRITISH STANDARDS INSTITUTION
    Design of concrete structures for retaining aqueous liquids
    BS 8007, 1987

10. AMERICAN CONCRETE INSTITUTE
    The effect of restraint, volume change and reinforcement on cracking of massive concrete
    ACI Manual of Concrete Practice, 207.2R-73, 1984

11. BAMFORTH, P B and GRACE, W R
    Early age thermal cracking in large sections – towards a design approach
    In: *Proceedings of Asia Pacific Conference in Roads, Highways and Bridges*
    Institute for International Research, Hong Kong, 1988

12. EMBORG, M
    Thermal stresses in concrete structures at early ages
    Doctoral Thesis, Lulea University of Technology, 1989

13 DANISH CONCRETE AND STRUCTURAL RESEARCH INSTITUTE
    CIMS - Computer integrated test and monitoring systems for construction sites
    CSRI, Dr Neergaards Vej 13, Postboks 82, DK-2970, Horsholm, Feb 1987

14. LANING, A
     Ocean-side foundation mat pour presents challenges
     Concrete Construction, Sept 1991, pp669-670

15. RANDALL, F A
     Concrete pumps complete massive foundation pour in 13½ hours
     Concrete Construction, Feb 1989, pp158-160

16   ANON
     Largest concrete pour
     ICI Bulletin, No. 32, Sept 1990

17. GRICE, J R and HEPPLEWHITE
     Design and construction of the Thames Barrier cofferdams
     Proc. Instn. Civ. Engrs, Part 1, Vol 74, May 1993, pp191-224

18. FITZGIBBON, M E
     Large pours - 3, continuous casting
     Concrete, Feb 1977, pp35-36

19. COOKE, T H
     Concrete pumping and spraying – a practical guide
     Thomas Telford, London, 1990

20. BRITISH CONCRETE PUMPING ASSOCIATION
     The manual and advisory safety code of practice for concrete pumping
     BCPA, Revised Edition, 1988

21. BRITISH STANDARDS INSTITUTION
     Concrete: methods for specifying concrete mixes
     BS 5328, Part 2, 1991

22. TEYCHENNE, D C *et al*
     Design of normal concrete mixes
     Dept. of Environment, 1988

23. AMERICAN CONCRETE INSTITUTE
     *Standard practice for selecting proportions for normal,
     heavyweight and mass concrete*
     ACI Manual of Concrete Practice, 211.1-81, 1984

24. BRITISH STANDARDS INSTITUTION
     Structural use of concrete: Part 1, Code of Practice for Design and Construction
     BS 8110, 1985

25. BABTIE, SHAW & MORTON
     Properties of concrete in relation to transportation, placing and finishing
     CIRIA Funders Report CP/18, in preparation

26. CONCRETE SOCIETY
     The use of ggbs and pfa in concrete
     Concrete Society, Technical Report, 40, 1991

27. RIXOM, M R and MAILVAGANAM, M P
     Chemical admixtures for concrete
     E & FN Spon, London, 1986

28. CLEAR, C A and HARRISON, T A
     Concrete pressure on formwork
     CIRIA Report 108, 1985

29. BRITISH STANDARDS INSTITUTION
Pulverised fuel ash, Part 1, Specification for pulverised fuel ash for use as a cementitious component in concrete
BS 3892: Part 1: 1982

30. BRITISH STANDARDS INSTITUTION
Specification for ground granulated blastfurnace slag for use with Portland cement
BS 6699: 1986

31. CONCRETE SOCIETY
Non structural cracks in concrete
Concrete Society, Technical Report 22, Third Ed, 1992

32. CONCRETE SOCIETY
The use of microsilica in concrete
Report of a Working Party, Technical Report No. 41, 1993

33. BAMFORTH, P B
Admixtures - A contractor's view
World of Concrete, Europe 86, London 1986

34. BAMFORTH, P B
In-situ measurement of the effect of partial Portland cement replacement using either fly ash or ground granulated blastfurnace slag on the performance of mass concrete
Proc. of Instn. of Civ Engrs, Part 2, Vol 69, Sept 1980, pp777-800

35. BAMFORTH, P B
An investigation into the influence of partial Portland cement replacement using either fly ash or ground granulated blastfurnace slag on the early age and long-term behaviour of concrete
Taywood Engineering Limited Research Report No. 914J/78/2067

36. LERCH, W and BOGUE, R H
Heat of hydration of Portland cement pastes
Journal of Res. Nat. Bur. Stand, Vol 12, No. 5, May 1934, pp645-664

37. CARLSON, R W, HOUGHTON, D L and POLIVKA, M
Causes and control of cracking in unreinforced mass concrete
ACI Journal, July 1979

38. COOLE, M J
Heat release characteristics of concrete containing ground granulated blastfurnace slag in simulated large pours
Mag. of Concr. Res., Vol. 40, No. 144, Set. 1988, pp152-158

39. BAMFORTH, P B
The effect of heat of hydration of pfa concrete and it effect on strength
ASHTECH 84, Second Int. Conf. on Ash Technology and Marketing
London, Sept 1984, pp287-294

40. OWENS, P L and BUTTLER, F G
The reactions of fly ash and Portland cement with relation to strength of concrete as a function of time and temperature
Proc. of 7th Int. Conf. on Chemistry of Cement, 3, IV-60
Paris 1980

41. DHIR, R K, MUNDAY, J G L and ONG, L T
Investigations of the engineering properties of OPC/pulverised fuel ash concrete:
Strength development and maturity
Proc. Inst. Civ. Engrs, Part 2, Vol 77, No. 6 June 1984, pp239-254.

42. CHINA LIGHT & POWER
pfa Concrete Studies, 1988-1998
Prepared by CLP, Taywood Engineering Limited and L G Mouchell and Ptnrs
(Asia), Vol, 9, Supplementary Report at Year 1 and Appendix, CLP 1991

43   COOLE, M J and HARRISON, A M
*The effect of simulated large pour curing on the temperature rise and strength
growth of pfa containing concrete*
Blue Circle Industries Plc, Research Division, Greenhithe, Kent

44. THOMAS, M D A and MATTHEWS, J D
Durability studies of pfa concrete structures
BRE Information Paper 11/91, June 1991

45. CONCRETE SOCIETY/I.STRUCT.E.
Formwork, a guide to good practice
Pub. Conc. Soc. 1986

46. REDDI, S A and APTE, L N
Production and placement of concrete of Trombay Power Station
Indian Concrete Journal, Sept. 1985, 232-249

47. HARRISON, T A
Formwork striking times - methods of assessment
CIRIA Report 73, 1987

48. BUNGEY, J G.
Testing concrete in structures
CIRIA Technical Note 143, 1992

49. BRITISH STANDARDS INSTITUTION
*Testing concrete – recommendations for the assessment of concrete strength by
near to surface tests*
BS 1881: Part 207, 1992

50. TURTON, C D
Private communication to P B Bamforth, March 1993

51. PRICE, W F and WIDDOWS, S J
The effects of permeable formwork on the surface properties of concrete
Magazine of Concrete Research, Vol. 43, No. 155, 93-104

52. AMERICAN CONCRETE INSTITUTE
Recommended practice for measuring, mixing and placing concrete
ACI, 304-73 (reaffirmed 1983)

53. AMERICAN CONCRETE INSTITUTE
Standard practice for consolidation of concrete
ACI, Manual of Concrete Practice 309-72 (Revised 1982), 1988

54. McCLELLAND, R A
Private communication to P B Bamforth, March 1993

55. BIRT, J C
Curing concrete – an appraisal of attitudes, practices
CIRIA Report 43 (Second Edition) 1981

56. CLEAR, C
    The effects of autogenous healing upon the leakage of water through cracks in concrete
    BCA Technical Report 559, 1985

57. ORCHARD, D F
    Concrete technology, Vol. 2 Practice
    Applied Science Publishers Ltd, London, 1973

58. FÉDÉRATION INTERNATIONALE DE LA PRÉCONTRAINTE
    Condensed silica fume in concrete
    FIP State-of-the-Art Report, Thomas Telford, 1988

59. AMERICAN CONCRETE INSTITUTE
    Cooling and insulating systems for mass concrete
    ACI Manual of Concrete Practice, 207.4R-80, 1988

60. KURITA, M, GOTO, S, MINEHISHI, K, NEGAMI, Y and KUWAHARA, T
    Precooling concrete using frozen sand
    Concrete International, June 1990, pp60-65

61. NAMBIAR, O N N and KRISHNAMURTHY, V
    Control of temperature in mass concrete pours
    Indian Concrete Journal, March 1984, pp67-73

62. ROBBINS, J
    Cool Customer
    New Civil Engineer, 12 Sept 1991

63. HENRIKSEN, K R
    Avoidance of cracking at construction joints and between solid sections
    Nordisk Betong 1- 1983, pp17-27

64. ANON
    West Germany – Sub surface rainwater reservoir
    Construction Industry International, Aug 1982, pp17-18

65. ALLEN, R T L and EDWARDS, S C (Eds)
    Repair of Concrete Structures
    Blackie and Sons Ltd, 1987

66. PULLAR-STRECKER, P
    Corrosion damaged concrete - assessment and repair
    CIRIA Book 1 (Published in conjunction with Butterworth Heinemann), London, 1987

67. BRITISH STANDARDS INSTITUTION
    Guide to assessment of concrete strength in existing structures
    BS 6089: 1981

68. ANON
    Tall order for SGB formwork
    Concrete, Jan/Feb, 1993, p41